This book is to be returned on or before
the last date stamped below.

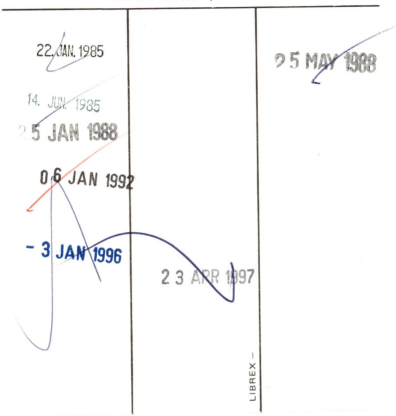

22. JAN. 1985

14. JUN. 1985

2 5 JAN 1988

0 6 JAN 1992

- 3 JAN 1996

2 3 APR 1997

2 5 MAY 1988

LIBREX —

Heat and Mass Transfer in Packed Beds

Topics in Chemical Engineering

A series of monographs and texts edited by R. Hughes, *University of Salford, UK.*

Volume 1 HEAT AND MASS TRANSFER IN PACKED BEDS
N. Wakao and S. Kaguei
Yokohama National University, Japan

Volume 2 THREE-PHASE CATALYTIC REACTORS
P. A. Ramachandran and R. V. Chaudhari
National Chemical Laboratory, Poona, India

Additional volumes in preparation

ISSN: 0277-5883

HEAT AND MASS TRANSFER IN PACKED BEDS

N. Wakao and S. Kaguei

Yokohama National University, Japan

GORDON AND BREACH SCIENCE PUBLISHERS
New York   London   Paris.

Copyright © 1982 by Gordon and Breach, Science Publishers, Inc.

Gordon and Breach, Science Publishers, Inc.
One Park Avenue
New York, NY 10016

Gordon and Breach Science Publishers Ltd.
42 William IV Street
London, WC2N 4DE

Gordon & Breach
58, rue Lhomond
75005 Paris

Library of Congress Cataloging in Publication Data

Wakao, Noriaki, 1930–
   Heat and mass transfer in packed beds.

   (Topics in chemical engineering, ISSN 0277-5883; v. 1)
   Includes bibliographies and indexes.
   1. Fluidization. 2. Heat – Transmission. 3. Mass transfer.
I. Kaguei, Seiichirō. II. Title. III. Series.
TP156.F65W34   660.2'842   81-13203
ISBN 0-677-05860-8   AACR2

Contents

Introduction to the Series

Chemical engineering covers a very wide spectrum of learning and the number of subject areas encompassed in both undergraduate and graduate courses is inevitably increasing every year. This wide variety of subjects makes it difficult to cover the whole subject matter of chemical engineering in a single book. The present series is therefore planned as a number of books covering areas of chemical engineering which, although important, are not treated at any length in graduate and postgraduate standard texts. Additionally, the series will incorporate recent research material which has reached the stage where an overall survey is appropriate, and where sufficient information is available to merit publication in book form for the benefit of the profession as a whole.

Inevitably, with a series such as this, constant revision is necessary if the value of the texts for both teaching and research purposes is to be maintained. I would therefore be indebted to individuals for criticisms and for suggestions for future editions.

R. HUGHES

Preface

Nori Wakao commenced his research in heat and mass transfer in packed bed reactors about two and a half decades ago when he was a graduate student. He was first interested in steady-state heat transfer in packed beds, particularly in axial and radial effective thermal conductivities. He was surprised to find from his experimental results that the axial effective thermal conductivities were larger than the radial conductivities, but soon learned that Wilhelm had also found, from tracer dispersion measurements, that the axial gas dispersion coefficient was larger than that in the radial direction.

A packed bed is a heterogeneous system composed of solid particles and fluid flowing in the interstitial space among the particles. Because of this heterogeneity and complexity, the packed bed has not lent itself to the application of exact hydrodynamic theory. However, instead of an exact theory, a rather conventional or statistical approach has often been made. One of the typical examples is the assumption made by Wilhelm that a packed bed may be regarded as a series of mixing cells, each containing a single particle. The development of the digital computer has made application of this idea possible for the analysis and design of packed bed reactors.

Ranz's model on fluid mixing in a sphere-lattice has also made us visualize where the lateral mixing comes from in a packed bed. His model was later extended by Yagi and Kunii for the interpretation and correlation of the radial effective thermal conductivities.

In packed bed heat transfer, what Gunn found from frequency thermal response measurements is of great significance. As far as the authors know, he was the first to observe the large axial fluid thermal diffusivities from the conventional model based on the assumption of the intraparticle temperature having radial symmetry. The interpretation of his finding was attempted separately by Vortmeyer and Wakao.

One of the long-lasting subjects of discussion in the past four decades has been the anomalous decrease in particle-to-fluid heat and mass transfer coefficients with decreasing flow rate at low Reynolds number. In fact, the anomaly had been experimentally observed by a number of investigators. The authors have shown from theory that fluid dispersion coefficients for mass depend upon the type of system: inert bed, reacting bed, or bed with only mass transfer taking place between particle surface and fluid. They have shown that the particle-to-fluid transfer coefficients never continue to decrease beyond a certain Reynolds number, if proper values of fluid dispersion coefficients are employed.

With regard to gaseous diffusion in porous media, the Wicke–Kallenbach type of apparatus is widely used for the determination of effective diffusivity. In the early work, there was some confusion in interpreting the effective diffusivity data in a binary gaseous system. It had been recognized that the inverse relation between the ratio of diffusion flux and the square root of the molecular weight of the gases in a binary gaseous system applied only to the Knudsen diffusion region. Hoogschagen, however, found experimentally that the same relation applied to the bulk diffusion as well. This important observation has led to the succeeding theoretical development made by Evans, Watson and Mason, Scott and Dullien, and Rothfeld on the gas diffusion in a capillary tube, covering the whole range from the bulk through the transitional stage to the Knudsen region. Pollard and Present made a theoretical computation of the gaseous self-diffusion coefficient so that they had to conduct the elaborate computation manually. But it is amazing that their results are in agreement with those predicted from a simple formula derived intuitively by Bosanquet. The importance of pore diffusion in catalysis was believed to have been first pointed out separately by Damköhler, Thiele and Zeldowitsch at about the same time in the late 1930s, but, in fact, a similar work was reported by Jüttner as early as in 1909. The work by Wheeler on his parallel pore model has contributed significantly to the theoretical development of diffusion in catalysis. Regretfully, despite all of these and since the pioneer work of Jüttner, research progress in the transport phenomena in porous catalysts has been rather slow.

The recently developed chromatography measurement techniques have been successfully applied for the determination of some of the rate parameters of interest in packed bed systems. Contributions toward the development of the various estimation techniques are made by Ostergaard and Michelsen, Anderssen and White, Smith, Silveston and Hudgins, and others.

In this preface, only the names of some researchers and their contributions are mentioned; however, the valuable contributions made by many other investigators should also be credited.

Lastly, the authors hope that this book will help graduate students and researchers in chemical engineering understand the phenomena of heat and mass transfer in packed bed reactors.

We wish to acknowledge the helpful assistance of Dr W. K. Teo and Dr R. Hughes in offering many constructive comments on the entire manuscript.

N. WAKAO
S. KAGUEI

Notation

A	area (m^2)
A_ω	$= A_\omega^{II}/A_\omega^{I}$, amplitude ratio for frequency ω
$A_\omega^{I}, A_\omega^{II}$	amplitudes for the harmonic components with frequency ω of the input and response signals, respectively: in Example 1.1 (s); in Chapters 6 and 7 (K)
a	particle surface area per unit volume of packed bed; $a = 3(1 - \epsilon_b)/R$ for spherical particles (m^{-1})
a	capillary tube radius (m)
$\bar{a}$	mean pore radius (m)
a_n	coefficient defined by Eq. (1.34a) (s^{-1}); defined by Eq. (1.41a) (mol m^{-3}); defined by Eq. (6.17a) (K)
a_n	n-th root of Eqs. (5.4c) or (8.6)
a_n^*	coefficient defined by Eq. (1.46a) (mol m^{-3})
a_n^+	coefficient defined by Eq. (1.42a) (mol m^{-3}); defined in Eq. (6.16) (K)
$a_\omega^{I}, a_\omega^{II}$	cosine components with frequency ω of input and response signals, respectively: in Section 1.1.5 (mol s m^{-3}); in Example 1.1 (s)
b_n	coefficient defined by Eq. (1.34b) (s^{-1}); defined by Eq. (1.41b) (mol m^{-3}); defined in Eq. (6.17b) (K)
b_n^*	coefficient defined by Eq. (1.46b) (mol m^{-3})
b_n^+	coefficient defined by Eq. (1.42b) (mol m^{-3}); defined in Eq. (6.16) (K)
$b_\omega^{I}, b_\omega^{II}$	sine components with frequency ω of input and response signals, respectively: in Section 1.1.5 (mol s m^{-3}); in Example 1.1 (s)
C	concentration in the bulk fluid (mol m^{-3})
C'	concentration in empty/inert bed sections (mol m^{-3})
C^{I}, C^{II}	input and response concentrations, respectively (mol m^{-3})
C^*	fluid concentration in a unit cell (mol m^{-3})
C^{*a}, C^{*s}	antisymmetric and symmetric components of C^*, respectively (mol m^{-3})
$C^*(II)$	fluid concentration in two cells in contact (mol m^{-3})

C_A, C_B concentrations of reacting species A and B, respectively (mol m^{-3})

C_{As}, C_{Bs} concentrations at the pellet surface of reacting species A and B, respectively (mol m^{-3})

C_{exit} exit concentration (mol m^{-3})

C_F specific heat of the fluid $(\text{J kg}^{-1}\,\text{K}^{-1})$

C_{in} inlet concentration (mol m^{-3})

C_{ps}, C_s concentrations at the particle surface (mol m^{-3})

C_S specific heat of the solid particle $(\text{J kg}^{-1}\,\text{K}^{-1})$

c gaseous tracer concentration in the intraparticle pore volume (mol m^{-3})

c^* intraparticle gaseous concentration in a unit cell (mol m^{-3})

c^{*a}, c^{*s} antisymmetric and symmetric components of c^*, respectively (mol m^{-3})

$c^*(\text{II})$ intraparticle gaseous concentration in two cells in contact (mol m^{-3})

c_a gaseous concentration in the macropores (mol m^{-3})

c_{ad} amount adsorbed in a particle (mol kg^{-1})

c_{ad}^* amount adsorbed in a unit cell (mol kg^{-1})

c_i gaseous concentration in microporous particle (mol m^{-3})

D' dispersion coefficient in empty bed sections $(\text{m}^2\,\text{s}^{-1})$

D_a gaseous effective diffusivity in the macropores, defined per unit cross-sectional area of the pellet $(\text{m}^2\,\text{s}^{-1})$

D_{ax} axial fluid dispersion coefficient based on unit void area $(\text{m}^2\,\text{s}^{-1})$

$(D_{ax})_{inert}$ axial fluid dispersion coefficient based on unit void area in bed of non-porous inert particles $(\text{m}^2\,\text{s}^{-1})$

$(D_{ax})_{mixing}$ turbulent contribution to D_{ax} $(\text{m}^2\,\text{s}^{-1})$

D_e intraparticle effective diffusivity $(\text{m}^2\,\text{s}^{-1})$

D_i gaseous effective diffusivity in the micropores, defined per unit cross-sectional area of the microporous particle $(\text{m}^2\,\text{s}^{-1})$

D_K Knudsen diffusivity $(\text{m}^2\,\text{s}^{-1})$

$(D_m)_{ext\,film}$ diffusivity of m-th component through external film on catalyst pellet $(\text{m}^2\,\text{s}^{-1})$

D_{mm} self-diffusion coefficient for species m $(\text{m}^2\,\text{s}^{-1})$

D_{mn} binary molecular diffusion coefficient for species m and n $(\text{m}^2\,\text{s}^{-1})$

D_p particle diameter (m)

D_r	radial fluid dispersion coefficient based on unit void area $(m^2\ s^{-1})$
$(D_r)_{inert}$	radial fluid dispersion coefficient based on unit void area in bed of non-porous inert particles $(m^2\ s^{-1})$
$(D_r)_{mixing}$	turbulent contribution to D_r $(m^2\ s^{-1})$
D_{si}	effective diffusivity of adsorbate in the micropores, defined per unit cross-sectional area of the microporous particle $(m^2\ s^{-1})$
D_T	column diameter (m)
D_v	molecular diffusion coefficient $(m^2\ s^{-1})$
E	activation energy of intrinsic chemical reaction $(J\ mol^{-1})$
E'	activation energy of overall reaction $(J\ mol^{-1})$
E°	effective diffusion coefficient in a cylindrical unit cell $(m^2\ s^{-1})$
$(E^\circ)_{inert}$	effective diffusion coefficient in a bed of non-porous inert particles $(m^2\ s^{-1})$
$E^\circ(II)$	effective diffusion coefficient in two cells in contact $(m^2\ s^{-1})$
E_{ax}	axial fluid dispersion coefficient based on column cross-section $(m^2\ s^{-1})$
E_f	catalyst effectiveness factor
$F(s)$	transfer function
$[F(s)]_{C-S}$	transfer function (C–S model)
$[F(s)]_{D-C}$	transfer function (D–C model)
$[F(s)]_{Schumann}$	transfer function (Schumann model)
F_{ij}	view factor from surface i to surface j
$\bar{F}_{ij}$	overall view factor from surface i to surface j
F_v	volumetric flow rate $(m^3\ s^{-1})$
f	radiation correction factor defined by Eq. (5.59)
$f(t)$	inversed transfer function, i.e. impulse response signal (s^{-1})
G	$= u\rho_F$, fluid mass velocity per unit area of bed cross-section $(kg\ m^{-2}\ s^{-1})$
h_p	particle-to-fluid heat transfer coefficient $(W\ m^{-2}\ K^{-1})$
$h_p^\dagger$	particle-to-fluid heat transfer coefficient evaluated with $\alpha_{ax} = 0$ $(W\ m^{-2}\ K^{-1})$
h_r	radiant heat transfer coefficient based on the unit area of the particle surface $(W\ m^{-2}\ K^{-1})$
h_r'	radiant heat transfer coefficient based on cross-sectional area $(W\ m^{-2}\ K^{-1})$

h_r^*	radiant heat transfer coefficient with radiating gray gas $(\text{W m}^{-2}\,\text{K}^{-1})$
h_w	wall heat transfer coefficient $(\text{W m}^{-2}\,\text{K}^{-1})$
I_n	imaginary part of $F(in\pi/\tau)$
$I_n(\ )$	modified Bessel function of the first kind and n-th order
I_ω	imaginary part of $F(i\omega)$
i	$=(-1)^{1/2}$
J	diffusion flux in a capillary tube $(\text{mol m}^{-2}\,\text{s}^{-1})$
J_e	diffusion flux in porous media $(\text{mol m}^{-2}\,\text{s}^{-1})$
J_{Heat}	$=Nu/(Pr^{1/3}Re)$, J factor for heat transfer
J_{Mass}	$=Sh/(Sc^{1/3}Re)$, J factor for mass transfer
$J_n(\ )$	Bessel function of the first kind and n-th order
K	overall rate constant (s^{-1})
K^o	effective thermal conductivity of cylindrical unit cell with stagnant fluid $(\text{W m}^{-1}\,\text{K}^{-1})$
K_A	adsorption equilibrium constant (first-order) $(\text{m}^3\,\text{kg}^{-1})$
K_{eq}	chemical reaction equilibrium constant
$K_n(\ )$	modified Bessel function of the second kind and n-th order
k_a	adsorption rate constant (first-order) $(\text{m}^3\,\text{kg}^{-1}\,\text{s}^{-1})$
k_e^o	effective thermal conductivity of a quiescent bed $(\text{W m}^{-1}\,\text{K}^{-1})$
$(k_e^o)_{\text{COND}}$	conduction contribution to k_e^o $(\text{W m}^{-1}\,\text{K}^{-1})$
$(k_e^o)_{\text{RAD-COND}}$	combined radiation and conduction contribution to k_e^o $(\text{W m}^{-1}\,\text{K}^{-1})$
$(k_e^o)_{\text{CONTACT}}$	contact contribution to k_e^o $(\text{W m}^{-1}\,\text{K}^{-1})$
k_{eax}	effective axial thermal conductivity $(\text{W m}^{-1}\,\text{K}^{-1})$
k_{eF}	effective fluid phase thermal conductivity, defined per unit area of bed cross-section (C–S model) $(\text{W m}^{-1}\,\text{K}^{-1})$
k_{er}	effective radial thermal conductivity $(\text{W m}^{-1}\,\text{K}^{-1})$
k_{eS}	effective solid phase thermal conductivity, defined per unit area of bed cross-section (C–S model) $(\text{W m}^{-1}\,\text{K}^{-1})$
k_F	fluid thermal conductivity $(\text{W m}^{-1}\,\text{K}^{-1})$
k_f	particle-to-fluid mass transfer coefficient (m s^{-1})
$k_f^\dagger$	particle-to-fluid mass transfer coefficient evaluated with $D_{\text{ax}}=0$ (m s^{-1})
k_S	solid thermal conductivity $(\text{W m}^{-1}\,\text{K}^{-1})$
k_x	chemical reaction rate constant (first-order) (s^{-1})
L	distance/height (m)

L_D	half length of dead volume section (m)
L_e	actual length of pore (m)
l	distance between a response measuring point and bed exit (m)
M	molecular weight
M	molar mass (kg mol^{-1})
M_n^I, M_n^{II}	n-th moments of input and response signals, respectively (s^n)
M_1^*	$= m_1^*/m_0^*$ (s)
m_n^{*I}, m_n^{*II}	n-th weighted moments of input and response signals, respectively (s^n)
N	diffusion rate (mol s^{-1})
N_D	$= D_{ax}/(LU)$, mass dispersion number
N_H	$= \alpha_{ax}/(LU)$, thermal dispersion number
Nu	$= h_p D_p/k_F$, Nusselt number (modified D–C model)
Nu'	Nusselt number (original D–C model)
Nu''	Nusselt number (C–S model)
Nu'''	Nusselt number (Schumann model)
Nu_r	$= h_r D_p/k_S$, radiant Nusselt number
n_d	molar flux from gas to pellet (mol m^{-2} s^{-1})
n_x	diffusion rate passing axially through a cross-sectional area of a unit cell (mol s^{-1})
P	total pressure (Pa)
Pe_{ax}	$= D_p U/D_{ax}$, axial Peclet number
$(Pe_{ax})_{mixing}$	$= D_p U/(D_{ax})_{mixing}$, turbulent contribution to Pe_{ax}
Pe_r	$= D_p U/D_r$, radial Peclet number
$(Pe_r)_{mixing}$	$= D_p U/(D_r)_{mixing}$, turbulent contribution to Pe_r
$P_n^m(\ )$	associated Legendre function of the first kind and n-th order
Pr	$= C_F \mu/k_F$, Prandtl number
P_0	atmospheric pressure (Pa)
p	emissivity of gray surface
p_g	emissivity of gray gas
Q	radiant heat transfer rate between two hemispheres (W)
Q_p	heat transfer rate from fluid to particle (W)
$\bar{q}$	average axial heat conduction rate in a cell (W)
$q_{lateral}$	lateral heat flow rate per unit area (W m^{-2})
q_v	rate of heat generation per unit volume of solid (W m^{-3})
q_x	axial heat conduction rate in a cell (W)

R	particle radius (m)
R'	cylindrical cell radius (m)
Re	$= D_p u \rho_F / \mu = D_p G / \mu$, Reynolds number
R_g	gas constant (J K^{-1} mol^{-1})
R_n	real part of $F(in\pi/\tau)$
R_p	total reaction rate in a single pellet (mol s^{-1})
R_T	column radius (m)
R_v	reaction rate based on stoichiometry (mol m^{-3} s^{-1})
R_ω	real part of $F(i\omega)$
r	radial distance variable from the center of a particle (m)
r	radial distance variable in a cylindrical packed bed (m)
r'	radial distance variable from central axis of a unit cell (m)
r_i	radial distance variable in microporous particle (m)
r_0	radius of microporous particle (m)
r_v	chemical reaction rate per unit volume of catalyst pellet (mol m^{-3} s^{-1})
r_x	reaction rate per unit volume of reactor (mol m^{-3} s^{-1})
Sc	$= \mu/(\rho_F D_v)$, Schmidt number
Sh	$= k_f D_p / D_v$, Sherwood number
$Sh^\dagger$	$= k_f^\dagger D_p / D_v$, Sherwood number evaluated with $D_{ax} = 0$
S_g	pore surface area per unit mass of porous solid (m^2 kg^{-1})
S_p	particle surface area (m^2)
s	Laplace operator (s^{-1})
T	temperature (K)
T_c	temperature at central axis (K)
T_{exit}	exit fluid temperature (K)
T_F	fluid temperature (K)
T_F^*	fluid temperature in a unit cell (K)
T_F^I, T_F^{II}	input and response temperature signals, respectively (K)
$(T_F^{II})_{C-S}$	response temperature signal (C–S model) (K)
$(T_F^{II})_{D-C}$	response temperature signal (D–C model) (K)
$(T_F^{II})'_{D-C}$	response temperature signal (original D–C model) (K)
$(T_F^{II})_{Schumann}$	response temperature signal (Schumann model) (K)
T_{in}	inlet fluid temperature (K)
T_L	bed exit temperature (K)
T_m	mixed mean temperature (K)
T_{ps}	temperature at particle surface (K)
T_S	solid temperature (K)
T_S^*	solid temperature in a unit cell (K)

T_W	wall temperature (K)
T_0	temperature at bed inlet (K)
t	time variable (s)
U	interstitial fluid velocity (m s^{-1})
U_0	overall heat transfer coefficient (W m^{-2} K^{-1})
u	superficial fluid velocity (m s^{-1})
u'	fluid velocity in empty section (m s^{-1})
V	reactor volume (m^3)
V_g	pore volume per unit mass of porous solid (m^3 kg^{-1})
V_p	particle volume (m^3)
$\bar{v}$	mean molecular velocity (m s^{-1})
X	conversion
x	axial distance variable (m)
$Y_n(\ )$	Bessel function of the second kind and n-th order
y_m	mole fraction of species m

Greek symbols

α	$= 1 + J_2/J_1$
α	accommodation coefficient
α	fraction defined by Eq. (5.25)
α_{ax}	axial fluid thermal dispersion coefficient (m^2 s^{-1})
α'_{ax}	axial fluid thermal dispersion coefficient defined by Eq. (6.3) (m^2 s^{-1})
α_F	$= k_F/(C_F \rho_F)$, fluid thermal diffusivity (m^2 s^{-1})
α_r	radial fluid thermal dispersion coefficient (m^2 s^{-1})
α_S	$= k_S/(C_S \rho_S)$, solid thermal diffusivity (m^2 s^{-1})
γ_0	defined by Eq. (6.19a)
γ_1	defined by Eq. (6.19b) (s)
Δ	difference defined by Eq. (1.55) (s)
δ	diffusibility
δ	temperature jump coefficient in Eq. (5.52) (m)
δ_H	coefficient defined by Eq. (6.21)
δ_0	defined by Eq. (1.63a)
δ_1	defined by Eq. (1.64a) (s)
ϵ	root-mean-square error
ϵ_a	macropore void fraction of a pellet
ϵ_b	bed void fraction
$\epsilon_f, \epsilon'_f, \epsilon''_f, \epsilon^*_f$	root-mean-square-errors (frequency response)

ϵ_i	micropore void fraction of pellet
ϵ_{ip}	volume fraction of a microporous particle in pellet
ϵ_p	intraparticle void fraction
ϵ_s	volume fraction of solid
ϵ_s	root-mean-square-error (step response)
θ	angle variable (rad)
Λ	fractional contact area
Λ_D	coefficient defined by Eq. (1.86a)
Λ_{ad}	coefficient defined by Eq. (1.90a)
Λ_H	coefficient defined by Eq. (6.38b)
λ	mean free path (m)
λ_0	mean free path at P_0 (m)
μ	fluid viscosity (Pa s)
μ_n^I, μ_n^{II}	n-th central moments of input and response signals, respectively (s^n)
μ_n^{*I}, μ_n^{*II}	n-th weighted central moments of input and response signals, respectively (s^n)
ρ_F	fluid density (kg m^{-3})
ρ_p, ρ_S	particle density (kg m^{-3})
σ	Stefan–Boltzmann constant (W m^{-2} K^{-4})
σ_H^2	variance defined by Eq. (6.19) (s^2)
σ_M^2	variance defined by Eq. (1.64) (s^2)
τ, τ^*	half period (s)
τ	tortuosity factor
$\bar{\tau}$	$= L/U$, mean residence time (s)
Φ	angle variable in spherical coordinates (rad)
ϕ	$= R(k_x/D_e)^{1/2}$ Jüttner modulus for first-order irreversible chemical reaction
ϕ_R	Jüttner modulus for first-order reversible chemical reaction, defined by Eq. (3.88)
ϕ_ω	$= \phi_\omega^{II} - \phi_\omega^I$, phase shift (rad)
$\phi_\omega^I, \phi_\omega^{II}$	phases for harmonic components with frequency ω of input and response signals, respectively (rad)
Ψ	pressure parameter defined by Eq. (5.57)
ω	frequency (rad s^{-1})

1 Parameter Estimation from Tracer Response Measurements

IN THIS chapter, the techniques of parameter estimation from the measurements of tracer input and response signals are discussed. A moment method was first proposed for the estimation of the packed bed parameters described by a dispersed plug flow model. The moment method is interesting in theory, but, in practice, its shortcomings are that tailing and the frontal portions of the signal are overly weighted in the evaluation of the moments. Response signals usually have long tails, and the experimental errors in the tailing portion, as well as truncation of the tailing portion give serious errors in the moment analysis.

To overcome this disadvantage, modified methods have been proposed. These include: a weighted moment method and transfer function fitting by Østergaard and Michelsen [1]; Fourier analysis by Gangwal *et al.* [2]; curve fitting by Clements [3] and others.

Clements suggested that accurate parameter determination should be made by curve fitting in the time or Laplace domain. Anderssen and White [4, 5] showed that if an optimum weighting factor was chosen, the weighted moment method was almost as good as the analysis in the time domain. A similar conclusion was also reached by Wolff *et al.* [6].

In general, the moment, weighted moment, transfer function fitting and Fourier analysis all deal with the measured signals multiplied by a time function called a weighting factor. It is easily understood that the best weighting factor is unity, i.e. the parameters are best determined by real-time analysis.

1.1 Dispersed Plug Flow of an Inert System

Consider an inert tracer imposed on a stream of fluid in a packed bed of non-porous particles. The concentration input signal, $C_{\text{expt}}^{\text{I}}(t)$, and the

1

concentration response signal, $C_{expt}^{II}(t)$, are measured at two downstream points, at a distance, L, apart in the bed. If the concentration, C, is uniform in the radial direction, the fundamental equation according to the dispersed plug flow model is

$$\frac{\partial C}{\partial t} = D_{ax} \frac{\partial^2 C}{\partial x^2} - U \frac{\partial C}{\partial x} \tag{1.1}$$

where D_{ax} is the axial fluid dispersion coefficient, U is the interstitial fluid velocity, and x is the axial distance. Assuming an infinite packed bed (for the criterion, see Section 1.4), and if the initial tracer concentration is zero throughout the bed, the transfer function of the bed within a distance L is expressed in terms of the measured signals, $C_{expt}^{I}(t)$ and $C_{expt}^{II}(t)$, by

$$F(s) = \frac{\displaystyle\int_0^\infty C_{expt}^{II} \exp(-st)\, dt}{\displaystyle\int_0^\infty C_{expt}^{I} \exp(-st)\, dt} \tag{1.2}$$

$$= \exp\left\{\frac{1}{2N_D}[1 - (1 + 4N_D \bar{\tau}s)^{1/2}]\right\} \tag{1.3}$$

where $\bar{\tau}$ and N_D are the mean residence time and mass dispersion number, respectively, defined as follows:

$$\bar{\tau} = \frac{L}{U} \tag{1.3a}$$

$$N_D = \frac{D_{ax}}{LU}. \tag{1.3b}$$

1.1.1   Moment Method for Impulse Response

If the input signal is a delta function, the denominator of Eq. (1.2) becomes $\int_0^\infty C_{expt}^{I}\, dt$, which is identical to $\int_0^\infty C_{expt}^{II}\, dt$. The transfer function is then

$$F(s) = \frac{\displaystyle\int_0^\infty C_{expt}^{II} \exp{(-st)}\, dt}{\displaystyle\int_0^\infty C_{expt}^{II}\, dt} \qquad (1.4)$$

Differentiating with respect to s, it gives

$$\frac{d^n F(s)}{ds^n} = (-1)^n \frac{\displaystyle\int_0^\infty C_{expt}^{II} t^n \exp{(-st)}\, dt}{\displaystyle\int_0^\infty C_{expt}^{II}\, dt} \qquad (1.5)$$

Therefore,

$$\left[\frac{d^n F(s)}{ds^n}\right]_{s=0} = (-1)^n M_n^{II} \qquad (1.6)$$

where

$$M_n^{II} = \frac{\displaystyle\int_0^\infty C_{expt}^{II} t^n\, dt}{\displaystyle\int_0^\infty C_{expt}^{II}\, dt} \qquad (1.7)$$

and is called the n-th moment of C_{expt}^{II}.

The second central moment or variance is defined as:

$$\mu_2^{II} = \frac{\displaystyle\int_0^\infty C_{expt}^{II}(t - M_1^{II})^2 \, dt}{\displaystyle\int_0^\infty C_{expt}^{II} \, dt} \tag{1.8}$$

$$= M_2^{II} - (M_1^{II})^2. \tag{1.9}$$

For the transfer function of Eq. (1.3), it is shown that

$$M_1^{II} = \bar{\tau} \tag{1.10}$$

$$\mu_2^{II} = 2(\bar{\tau})^2 N_D. \tag{1.11}$$

The mean residence time, $\bar{\tau}$, and the mass dispersion number, N_D, are determined, therefore, from the first moment and the variance of the impulse response, respectively.

1.1.2   Moment Method for One-Shot Input

Mathematically, delta input is the only possibility. Even if a tracer is injected instantaneously, the tracer is usually being imposed on a fluid flowing in a column, which results in some diffusion from the very beginning.

If tracer concentration–time curves are measured at two downstream points, the tracer can be introduced in any type of one-shot input. It is advantageous not to have to be concerned about the shape of the concentration curve on introducing the tracer.

Some mathematical manipulations give:

$$M_1^{II} - M_1^I = \bar{\tau} \tag{1.12}$$

$$\mu_2^{II} - \mu_2^I = 2(\bar{\tau})^2 N_D \tag{1.13}$$

where M_1^I and μ_2^I are the first and second central moments of input signal, respectively, and are defined in a similar way to the response signal moments in Eqs. (1.7) and (1.8). When the input signal is of delta function, $M_1^1 = 0$ and $\mu_2^1 = 0$, so that Eqs. (1.12) and (1.13) reduce, correspondingly, to Eqs. (1.10) and (1.11).

However, the moment method has the shortcoming that the weighting factor, t^n, puts a large weight on the tailing portion of the signal. Errors in the tailing portion are magnified in the evaluation of moments, particularly of the higher moments. Also the errors in the frontal portion are magnified in evaluating the central moments.

1.1.3 Weighted Moment Method

To overcome the shortcomings of the moment method, Østergaard and Michelsen [1] proposed using the form of the right-hand side of Eq. (1.5) as the basis of their analysis. This modified technique is called the weighted moment method. The weighting factor, $t^n \exp(-st)$, which is zero at both $t = 0$ and at longer times, may obviously avoid the disadvantages of the moment method. The parameters involved in the transfer function are then determined from the modified moments as defined below:†

Zeroth weighted moment

$$m_0^* = \frac{\displaystyle\int_0^\infty C \exp(-st)\, dt}{\displaystyle\int_0^\infty C\, dt} \tag{1.14}$$

n-th weighted moment

$$m_n^* = \frac{\displaystyle\int_0^\infty C t^n \exp(-st)\, dt}{\displaystyle\int_0^\infty C\, dt} \tag{1.15}$$

† Note that Østergaard and Michelsen [1] defined m_n^*/m_0^* as the n-th weighted moment.

n-th weighted central moment is defined in terms of $M_1^* = m_1^*/m_0^*$ as

$$\mu_n^* = \frac{\int_0^\infty C(t - M_1^*)^n \exp(-st)\,dt}{\int_0^\infty C\,dt} \tag{1.16}$$

The weighted moments are related to the transfer function as follows:

$$\frac{m_0^{*\mathrm{II}}}{m_0^{*\mathrm{I}}} = F(s) \tag{1.17}$$

$$\frac{m_1^{*\mathrm{II}}}{m_0^{*\mathrm{II}}} - \frac{m_1^{*\mathrm{I}}}{m_0^{*\mathrm{I}}} = -\frac{F'(s)}{F(s)} \tag{1.18}$$

$$\frac{\mu_2^{*\mathrm{II}}}{m_0^{*\mathrm{II}}} - \frac{\mu_2^{*\mathrm{I}}}{m_0^{*\mathrm{I}}} = \frac{d}{ds}\left[\frac{F'(s)}{F(s)}\right] \tag{1.19}$$

$$\frac{\mu_3^{*\mathrm{II}}}{m_0^{*\mathrm{II}}} - \frac{\mu_3^{*\mathrm{I}}}{m_0^{*\mathrm{I}}} = -\frac{d^2}{ds^2}\left[\frac{F'(s)}{F(s)}\right] \tag{1.20}$$

$$\frac{\mu_4^{*\mathrm{II}}}{m_0^{*\mathrm{II}}} - \frac{\mu_4^{*\mathrm{I}}}{m_0^{*\mathrm{I}}} - 3\left[\left(\frac{\mu_2^{*\mathrm{II}}}{m_0^{*\mathrm{II}}}\right)^2 - \left(\frac{\mu_2^{*\mathrm{I}}}{m_0^{*\mathrm{I}}}\right)^2\right] = \frac{d^3}{ds^3}\left[\frac{F'(s)}{F(s)}\right] \tag{1.21}$$

For the transfer function given by Eq. (1.3), it is shown that

$$\frac{F'(s)}{F(s)} = -\bar{\tau}(1 + 4N_\mathrm{D}\bar{\tau}s)^{-1/2} \tag{1.22}$$

$$\frac{d^n}{ds^n}\left[\frac{F'(s)}{F(s)}\right] = \frac{(2n)!}{n!}(-\bar{\tau})^{n+i}N_\mathrm{D}^n(1 + 4N_\mathrm{D}\bar{\tau}s)^{-n-1/2}. \tag{1.23}$$

The two parameters, $\bar{\tau}$ and N_D, can hence be determined from any set of two moment equations, for example, from Eqs. (1.17) and (1.18), or from Eqs. (1.18) and (1.19).

However, the problem is what optimal values of s to use in the parameter estimation; the weighting factors, $\exp(-st)$ for zeroth moment, $t\exp(-st)$ for first moment, $(t-M_1^*)^2\exp(-st)$ for second central moment etc. all give weight on different portions of the signal curve. In other words, the optimal value of s depends on the order of the moment.

Anderssen and White [5] have suggested the following equation for the estimation of the optimal s values:

$$s = \frac{n_{highest}}{t_{max\text{-}input} + t_{max\text{-}response} - \Delta t_D} \qquad (1.24)$$

where $n_{highest}$ is the highest order of moment used for the parameter estimation, Δt_D is the difference in time delay between the input and response signals, and t_{max} is the time when a signal reaches the maximum point.

1.1.4   Transfer Function Fitting

The transfer function of Eq. (1.3) may be rewritten as:

$$-[\ln F(s)]^{-1} = \bar{\tau}s[\ln F(s)]^{-2} - N_D \qquad (1.25)$$

or

$$\left[\frac{F'(s)}{F(s)}\right]^{-2} = \frac{4N_D}{\bar{\tau}}s + (\bar{\tau})^{-2}. \qquad (1.26)$$

Østergaard and Michelsen [1] recommended that the transfer function, evaluated from the measured input and response signals, should be plotted as $-[\ln F(s)]^{-1}$ versus $s[\ln F(s)]^{-2}$, or $[F'(s)/F(s)]^{-2}$ versus s. The parameters $\bar{\tau}$ and N_D are thus obtained from the slope and intercept of a straight line, according to Eq. (1.25) or Eq. (1.26).

Again, the problem is the selection of s values for the plot. If s is too large, the large weight is given to the frontal portion of the signal, while Eq. (1.3) shows that if s is too small, the transfer function itself has little dispersion effect. Hopkins *et al.* [7] recommended that transfer function fitting should be made within the restricted range $2 \leqslant s\bar{\tau} \leqslant 5$.

1.1.5   Fourier Analysis

Gangwal *et al.* [2] applied Fourier analysis to the estimation of parameters in adsorption chromatography.

Input and response signals may be considered to be composed of numerous harmonic components. Fourier analysis evaluates decay in amplitude and the phase shift for the harmonic components between the input and response signals.

Substitution of $s = i\omega$ (where ω is frequency) into a transfer function gives

$$F(i\omega) = R_\omega + iI_\omega. \tag{1.27}$$

For $F(s)$ according to Eq. (1.3), R_ω and I_ω are as follows:

$$
\begin{aligned}
R_\omega &= \exp\,(y)\cos z \\
I_\omega &= -\exp\,(y)\sin z
\end{aligned}
\tag{1.28}
$$

where

$$y = \frac{1}{2N_D} - a\cos b \tag{1.28a}$$

$$z = a\sin b \tag{1.28b}$$

$$a = \left[\left(\frac{1}{2N_D}\right)^4 + \left(\frac{\bar{\tau}\omega}{N_D}\right)^2\right]^{1/4} \tag{1.28c}$$

and

$$b = \tfrac{1}{2}\tan^{-1}(4N_D\bar{\tau}\omega). \tag{1.28d}$$

From the input and response signals measured, $F(i\omega)$ is evaluated from

$$F(i\omega) = \frac{\displaystyle\int_0^\infty C_{\mathrm{expt}}^{\mathrm{II}}\exp\,(-i\omega t)\,\mathrm{d}t}{\displaystyle\int_0^\infty C_{\mathrm{expt}}^{\mathrm{I}}\exp\,(-i\omega t)\,\mathrm{d}t} \tag{1.29}$$

$$= \frac{a_\omega^{II} - ib_\omega^{II}}{a_\omega^{I} - ib_\omega^{I}} \tag{1.30}$$

where

$$a_\omega^{I} = \int_0^\infty C_{expt}^{I} \cos \omega t \, dt \tag{1.30a}$$

$$b_\omega^{I} = \int_0^\infty C_{expt}^{I} \sin \omega t \, dt \tag{1.30b}$$

$$a_\omega^{II} = \int_0^\infty C_{expt}^{II} \cos \omega t \, dt \tag{1.30c}$$

$$b_\omega^{II} = \int_0^\infty C_{expt}^{II} \sin \omega t \, dt. \tag{1.30d}$$

By equating Eqs. (1.27) and (1.30), the parameters involved in the real and imaginary parts of Eq. (1.28) are determined in terms of the Fourier coefficients evaluated from the signals measured.

1.1.6   Curve Fitting in the Time Domain

This is a method in which the response signals measured are compared in the time domain with those predicted based on assumed parameter values. If the two signal curves agree well, the parameter values used for the prediction may be regarded as correct. The following methods may be applied.

1.1.6.1   *Prediction of the signal in response to one-shot input by a convolution integral*

Equation (1.2) indicates that the response signal is calculated by a convolution integral as:

$$C^{II}_{calc}(t) = \int_0^t C^{I}_{expt}(\xi) f(t - \xi)\, d\xi \tag{1.31}$$

where $f(t)$, the impulse response of a delta input, is the Laplace inversion of the transfer function defined as:

$$f(t) = \mathcal{L}^{-1}[F(s)] \tag{1.32}$$

or

$$\int_0^{\infty} f(t) \exp(-st)\, dt = F(s). \tag{1.32a}$$

In the case of dispersed plug flow of an inert fluid, $f(t)$ is easily found from Eq. (1.3) to be

$$f(t) = \frac{1}{2\bar{\tau}\left[\pi N_D \left(\dfrac{t}{\bar{\tau}}\right)^3\right]^{1/2}} \exp\left[-\frac{\left(1 - \dfrac{t}{\bar{\tau}}\right)^2}{4N_D \dfrac{t}{\bar{\tau}}}\right]. \tag{1.33}$$

However, in many cases, the transfer functions are much more complicated so that the inversion cannot be made easily by conventional methods. Under such conditions, the inversion has to be performed in terms of a Fourier series.

Over the period, 0 to 2τ, where 2τ is a period of time sufficiently long enough for the tailing portion of the response signal to vanish, $f(t)$ is expressed as:

$$f(t) = \frac{a_0}{2} + \sum_{n=1}^{\infty} \left(a_n \cos\frac{n\pi t}{\tau} + b_n \sin\frac{n\pi t}{\tau}\right) \tag{1.34}$$

where

$$a_n = \frac{1}{\tau} \int_0^{2\tau} f(t) \cos \frac{n\pi t}{\tau} \, dt \qquad (1.34\text{a})$$

$$b_n = \frac{1}{\tau} \int_0^{2\tau} f(t) \sin \frac{n\pi t}{\tau} \, dt. \qquad (1.34\text{b})$$

On the other hand, substitution of $s = in\pi/\tau$ into Eq. (1.32a) gives

$$\int_0^\infty f(t) \cos \frac{n\pi t}{\tau} \, dt = \text{Real} \left[F \left(i \frac{n\pi}{\tau} \right) \right]$$

$$= R_n \qquad (1.35)$$

and

$$-\int_0^\infty f(t) \sin \frac{n\pi t}{\tau} \, dt = \text{Imag} \left[F \left(i \frac{n\pi}{\tau} \right) \right]$$

$$= I_n. \qquad (1.36)$$

Note that in the present case of dispersed plug flow of an inert fluid, $F(0) = 1$, so that

$$R_0 = 1. \qquad (1.37)$$

The signal, in response to an imperfect pulse input, is zero at $t = 2\tau$, so that the response to a perfect delta input, or the impulse response $f(t)$, must become zero at a time shorter than 2τ. The terms on the left hand sides of Eqs. (1.35) and (1.36) may then be integrated from 0 to 2τ instead of 0 to ∞, and Eq. (1.34) is rewritten in terms of R_n and I_n evaluated from the transfer function as:

$$f(t) = \frac{1}{2\tau} + \frac{1}{\tau} \sum_{n=1}^\infty \left(R_n \cos \frac{n\pi t}{\tau} - I_n \sin \frac{n\pi t}{\tau} \right). \qquad (1.38)$$

Equation (1.38) can also be derived more directly from the following inversion integral:

$$f(t) = \frac{1}{2\pi} \int_{-\infty}^{\infty} F(i\omega) \exp(i\omega t)\, d\omega. \tag{1.39}$$

The integration is rewritten as:

$$f(t) = \frac{\Delta\omega}{2\pi} \sum_{n=-\infty}^{\infty} F(in\Delta\omega) \exp(in\Delta\omega t)$$

$$= \frac{\Delta\omega}{2\pi} \left\{ F(0) + \sum_{n=1}^{\infty} [F(in\Delta\omega) \exp(in\Delta\omega t) \right.$$

$$\left. + F(-in\Delta\omega) \exp(-in\Delta\omega t)] \right\}$$

$$= \frac{\Delta\omega}{2\pi} \left\{ F(0) + 2 \sum_{n=1}^{\infty} [R_n \cos(n\Delta\omega t) \right.$$

$$\left. - I_n \sin(n\Delta\omega t)] \right\}. \tag{1.40}$$

By considering $F(0) = 1$ and writing $\Delta\omega = \pi/\tau$, it is easily shown that Eq. (1.40) reduces to Eq. (1.38).

The response curve, $C_{\text{calc}}^{\text{II}}(t)$, is then computed from Eqs. (1.31) and (1.38) and may be compared with the experimental curve, $C_{\text{expt}}^{\text{II}}(t)$. The comparison is made over the entire region or any arbitrary time interval.

1.1.6.2   *Prediction of the signal in response to one-shot input by Fourier series*

The input signal measured is expressed by a Fourier series as:

$$C_{\text{expt}}^{\text{I}}(t) = \frac{a_0}{2} + \sum_{n=1}^{\infty} \left(a_n \cos \frac{n\pi t}{\tau} + b_n \sin \frac{n\pi t}{\tau} \right) \tag{1.41}$$

with the Fourier coefficients evaluated by the following expressions:

$$a_n = \frac{1}{\tau} \int_0^{2\tau} C^I_{\text{expt}} \cos \frac{n\pi t}{\tau} \, dt \qquad (1.41\text{a})$$

and

$$b_n = \frac{1}{\tau} \int_0^{2\tau} C^I_{\text{expt}} \sin \frac{n\pi t}{\tau} \, dt \qquad (1.41\text{b})$$

where 2τ is again a period of time long enough to let the tail of the response signal vanish.

The response signal is also predicted by a Fourier series of the form:

$$C^{II}_{\text{calc}}(t) = \frac{a_0^\dagger}{2} + \sum_{n=1}^{\infty} \left(a_n^\dagger \cos \frac{n\pi t}{\tau} + b_n^\dagger \sin \frac{n\pi t}{\tau} \right) \qquad (1.42)$$

where the Fourier coefficients, expressed in terms of $C^{II}_{\text{calc}}(t)$, are

$$a_n^\dagger = \frac{1}{\tau} \int_0^{2\tau} C^{II}_{\text{calc}} \cos \frac{n\pi t}{\tau} \, dt \qquad (1.42\text{a})$$

and

$$b_n^\dagger = \frac{1}{\tau} \int_0^{2\tau} C^{II}_{\text{calc}} \sin \frac{n\pi t}{\tau} \, dt. \qquad (1.42\text{b})$$

The transfer function can be written as:

$$F(s) = \frac{\displaystyle\int_0^{\infty} C^{II}_{\text{calc}} \exp(-st) \, dt}{\displaystyle\int_0^{\infty} C^I_{\text{expt}} \exp(-st) \, dt}. \qquad (1.43)$$

Substitution of $s = in\pi/\tau$ and consideration of the response signal being zero at $t \geqslant 2\tau$ give

$$F\left(i\,\frac{n\pi}{\tau}\right) = \frac{\displaystyle\int_0^{2\tau} C_{\text{calc}}^{\text{II}} \exp\left(-in\pi t/\tau\right) \, \mathrm{d}t}{\displaystyle\int_0^{2\tau} C_{\text{expt}}^{\text{I}} \exp\left(-in\pi t/\tau\right) \, \mathrm{d}t} \cdot \qquad (1.44)$$

From Eqs. (1.41a), (1.41b), (1.42a), (1.42b) and (1.44), it is shown that

$$a_n^\dagger - ib_n^\dagger = (a_n - ib_n)\, F\left(i\,\frac{n\pi}{\tau}\right). \qquad (1.45)$$

The Fourier coefficients in Eq. (1.42) are evaluated, therefore, with the measured input signal and the transfer function.

Similarly, the response signal measured is also expressed as:

$$C_{\text{expt}}^{\text{II}}(t) = \frac{a_0^*}{2} + \sum_{n=1}^{\infty} \left(a_n^* \cos\frac{n\pi t}{\tau} + b_n^* \sin\frac{n\pi t}{\tau}\right) \qquad (1.46)$$

with

$$a_n^* = \frac{1}{\tau}\int_0^{2\tau} C_{\text{expt}}^{\text{II}} \cos\frac{n\pi t}{\tau}\, \mathrm{d}t \qquad (1.46a)$$

and

$$b_n^* = \frac{1}{\tau}\int_0^{2\tau} C_{\text{expt}}^{\text{II}} \sin\frac{n\pi t}{\tau}\, \mathrm{d}t. \qquad (1.46b)$$

The root-mean-square-error between $C_{\text{expt}}^{\text{II}}$ and $C_{\text{calc}}^{\text{II}}$ over the entire region is then evaluated by Eq. (1.47) or Eq. (1.48).

$$\epsilon = \left[\frac{\displaystyle\int_0^{2\tau} (C_{\text{expt}}^{\text{II}} - C_{\text{calc}}^{\text{II}})^2 \, dt}{\displaystyle\int_0^{2\tau} (C_{\text{expt}}^{\text{II}})^2 \, dt} \right]^{1/2} \tag{1.47}$$

$$= \left\{ \frac{2\left(\dfrac{a_0^*}{2} - \dfrac{a_0^\dagger}{2}\right)^2 + \displaystyle\sum_{n=1}^{\infty} [(a_n^* - a_n^\dagger)^2 + (b_n^* - b_n^\dagger)^2]}{2\left(\dfrac{a_0^*}{2}\right)^2 + \displaystyle\sum_{n=1}^{\infty} (a_n^{*2} + b_n^{*2})} \right\}^{1/2} . \tag{1.48}$$

Note that as far as the tracer imposed is not dispersed in the column (inert or reversible adsorption system), $a_0^* = a_0^\dagger$.

In the evaluation of the confidence range by curve fitting in the time domain, the following general criteria are often adopted: fitting is good if $\epsilon < 0.05$; fitting is poor if $\epsilon \geqslant 0.05$.

The one-shot input method, discussed in Sections 1.1.6.1 and 1.1.6.2, requires that the tracer be imposed on a stream of an inert carrier fluid under the condition of $C = 0$ at $t \leqslant 0$. However, even if the measured input signal is any arbitrary function of time, without the imposed restriction, the signal in response to it can also be predicted by using a convolution integral. Let us assume that the impulse response, $f(t)$, becomes zero at any time greater than τ'. Equation (1.31) is then rewritten as:

$$C_{\text{calc}}^{\text{II}}(t) = \int_{t-\tau'}^{t} C_{\text{expt}}^{\text{I}}(\xi) f(t - \xi) \, d\xi. \tag{1.49}$$

Therefore, if the input signal is measured over a time interval from t_1 to t_2 (with the restriction that $t_2 - t_1 > \tau'$), the response signal in the range from $t_1 + \tau'$ to t_2 can be predicted from Eq. (1.49). The computed response signal, $C_{\text{calc}}^{\text{II}}(t)$, is then compared with the measured signal, $C_{\text{expt}}^{\text{II}}(t)$, in the time domain.

Example 1.1

Table 1.1 lists the input and response signal readings for nitrogen, an inert tracer imposed on a laminar flow of hydrogen in a packed bed

TABLE 1.1
Measured input and response signals for nitrogen.

$t(s)$	Input	Response	$t(s)$	Input	Response
0.0	0.0	—	16.5	4.0	22.5
0.5	9.6	—	17.0	3.5	19.5
1.0	40.6	—	17.5	3.3	17.0
1.5	84.5	—	18.0	3.0	14.5
2.0	116.7	—	18.5	2.7	12.8
2.5	131.6	—	19.0	2.3	11.0
3.0	134.1	—	19.5	2.1	9.8
3.5	129.0	—			
4.0	119.5	—	20.0	1.8	8.5
4.5	106.6	—	20.5	1.7	7.5
			21.0	1.5	6.5
5.0	95.3	0.0	21.5	1.4	5.5
5.5	82.2	2.3	22.0	1.3	4.5
6.0	70.1	11.2	22.5	1.2	4.0
6.5	61.3	21.0	23.0	1.1	3.5
7.0	52.6	35.0	23.5	1.0	3.1
7.5	45.0	48.6	24.0	0.9	2.7
8.0	39.2	61.5	24.5	0.8	2.4
8.5	33.8	70.7			
9.0	28.5	79.0	25.0	0.7	2.0
9.5	25.0	82.5	25.5	0.6	1.7
			26.0	0.5	1.4
10.0	21.9	83.0	26.5	0.4	1.3
10.5	19.0	83.0	27.0	0.3	1.2
11.0	16.5	80.5	27.5	0.2	1.1
11.5	14.5	76.0	28.0	0.1	1.0
12.0	12.3	70.0	28.5	0.1	0.9
12.5	11.0	62.8	29.0	0.0	0.8
13.0	9.5	55.8	29.5	—	0.7
13.5	8.4	49.7			
14.0	7.3	43.5	30.0	—	0.6
14.5	6.6	38.6	30.5	—	0.5
			31.0	—	0.4
15.0	5.8	33.7	31.5	—	0.4
15.5	5.2	29.6	32.0	—	0.2
16.0	4.5	25.5	32.5	—	0.1
			33.0	—	0.0

($L = 10.7$ cm and $\epsilon_b = 0.4$) of glass beads (at 20°C and atmospheric pressure). The transfer function is expressed by Eq. (1.3). Find $\bar{\tau}$ and N_D.

SOLUTION

Moment method

The input and response signals normalized by Eq. (1.50) are shown in Figure 1.1(a) (note that C^I and C^{II} are dimensionless).

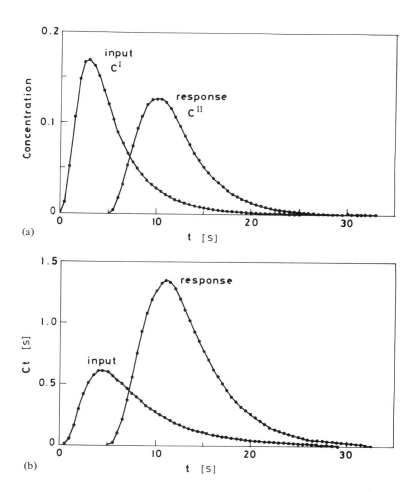

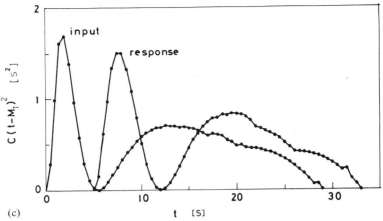

FIGURE 1.1   Curves for moment calculation for Example 1.1: (a) normalized input and response signals; (b) Ct for first moment; (c) $C(t-M_1)^2$ for variance.

$$\int_{0}^{\infty} C^{\mathrm{I}}\, dt = \int_{0}^{\infty} C^{\mathrm{II}}\, dt = 1 \text{ s.} \qquad (1.50)$$

The first moments are the areas under the curves, Ct versus t, in Figure 1.1(b). The areas give $M_1^{\mathrm{I}} = 5.4$ s and $M_1^{\mathrm{II}} = 12.0$ s, and consequently, $\bar{\tau} = 6.6$ s.

Similarly, the variances are the areas under the curves, $C(t-M_1)^2$ versus t, in Figure 1.1(c). The zig-zag curves obtained from the slightly scattered data at long periods of time are apparently the result of small errors in the tailing portions of the signals. This shows that the variance values themselves may have considerable errors. In any case, the difference between these two variances gives $N_{\mathrm{D}} = 0.0087$.

Weighted moment method
The weighting factors for the zeroth, first, second and third weighted moments are shown in Figures 1.2(a)-(d).

Figures 1.3(a)-(d) show $C\exp(-st)$, $Ct\exp(-st)$, $C(t-M_1^*)^2\exp(-st)$ and $C(t-M_1^*)^3\exp(-st)$ versus t curves. The zeroth, first and second weighted central moments are the areas under the curves of Figures 1.3(a), (b) and (c), respectively. In Figure 1.3(d), the curve is negative for $t < M_1^*$

and becomes positive for $t > M_1^*$. The third central moment is then the difference between the area of the positive portion of the curve $(t > M_1^*)$ and that of the negative portion $(t < M_1^*)$.

In Figure 1.3(c), the tailing portions are again zig-zag when s is small, but apparently become smooth with an increase in s. The same trend is seen in Figure 1.3(d). This is due to the fact that a large value of s makes the weight shift from the tailing portion to the frontal portion. The small errors associated with the frontal portions are then magnified when the moments are evaluated with large s values.

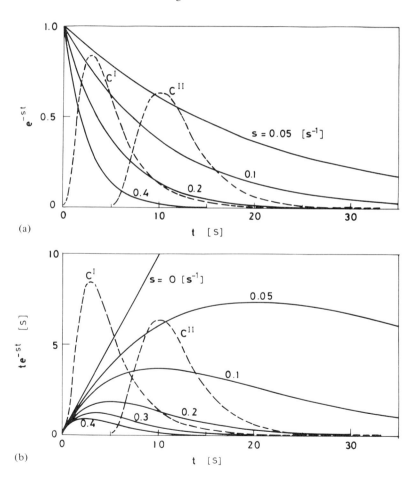

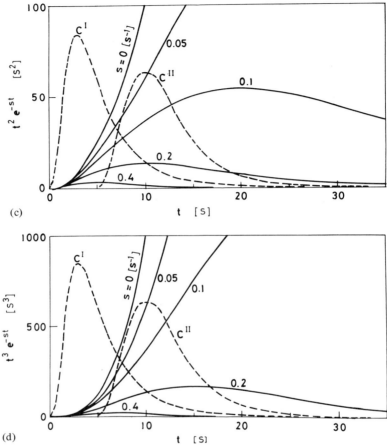

FIGURE 1.2   Weighting factors versus t for weighted moment calculation: (a) $\exp(-st)$ for zeroth weighted moment; (b) $t \exp(-st)$ for first weighted moment; (c) $t^2 \exp(-st)$ for second weighted moment; (d) $t^3 \exp(-st)$ for third weighted moment.

Table 1.2 lists the parameter values obtained in the range $s = 0.01$ to 2 s^{-1} from: (i) Eqs. (1.17) and (1.18), (ii) Eqs. (1.18) and (1.19), and (iii) Eqs. (1.19) and (1.20). Compared with the values of $\bar{\tau}$ and N_D obtained in (i) and (ii), the data determined in (iii) are entirely inconsistent and erroneous. Some of the parameter values are found to be negative or even imaginary. This discrepancy is obviously due to the fact that the third moments themselves, as seen in Figure 1.3(d), have large errors. As shown in Figure 1.4, the values of $\bar{\tau}$ obtained in (i) and (ii) in the range $s < 0.4\,\mathrm{s}^{-1}$ are slightly s-dependent, and they decrease at higher s values. The N_D values obtained in (i) and (ii) are, as shown in Figure 1.5, highly s-dependent. First, they increase with an increase in s, and then decrease after reaching maximum values.

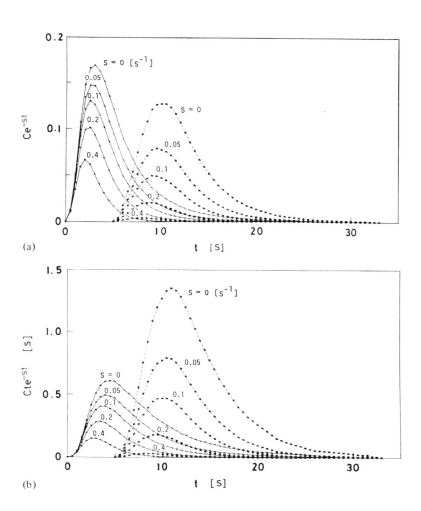

(a)

(b)

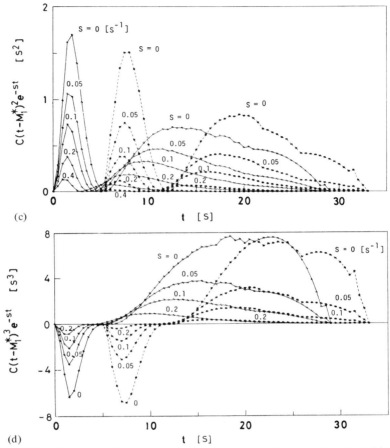

FIGURE 1.3   Curves for weighted moment calculation for Example 1.1 (solid and dashed lines are for input and response signals, respectively): (a) $C \exp(-st)$ for zeroth weighted moment; (b) $Ct \exp(-st)$ for first weighted moment; (c) $C(t - M_1^*)^2 \exp(-st)$ for second weighted central moment; (d) $C(t - M_1^*)^3 \exp(-st)$ for third weighted central moment.

From Figure 1.2, the values of s suited for both input and response curves seem to be

$$s \lesssim 0.1 \text{ s}^{-1} \qquad \text{for zeroth weighted moment}$$
$$s = 0.1 \text{ to } 0.2 \text{ s}^{-1} \qquad \text{for first weighted moment}$$
$$s = 0.2 \text{ to } 0.3 \text{ s}^{-1} \qquad \text{for second weighted moment.}$$

TABLE 1.2
The values of $\bar{\tau}$ and N_D obtained from the weighted moment method.

s (s^{-1})	(i) 0th and 1st moments		(ii) 0th, 1st and 2nd moments		(iii) 0th, 2nd and 3rd moments	
	$\bar{\tau}$ (s)	N_D	$\bar{\tau}$ (s)	N_D	$\bar{\tau}$ (s)	N_D
0.01	6.6	0.011	6.6	0.011	−0.13	24.4
0.02	6.6	0.012	6.6	0.014	−0.22	9.7
0.04	6.6	0.015	6.6	0.017	−0.49	2.3
0.06	6.6	0.017	6.6	0.020	−0.98	0.68
0.08	6.6	0.019	6.7	0.022	−1.9	0.21
0.1	6.6	0.020	6.7	0.023	−3.9	0.055
0.2	6.7	0.024	6.7	0.025	6.1	0.031
0.3	6.7	0.024	6.7	0.024	4.0	0.085
0.4	6.7	0.024	6.6	0.022	3.6	0.12
0.6	6.6	0.022	6.4	0.018	3.2	0.22
0.8	6.6	0.020	6.3	0.014	3.5	0.42
1.0	6.5	0.017	6.1	0.011	_a	_a
2.0	6.1	0.009	5.4	0.003	_a	_a

[a] Imaginary values obtained.

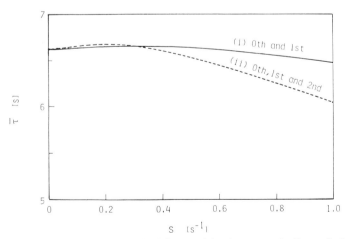

FIGURE 1.4 $\bar{\tau}$ versus s, obtained from weighted moment for Example 1.1.

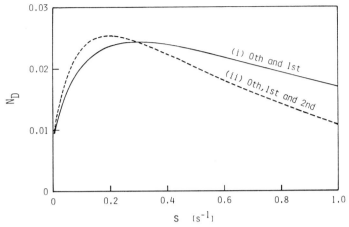

FIGURE 1.5 N_D versus s, obtained from weighted moment for Example 1.1.

If we roughly assume that $s = 0.1$ s^{-1} is good for the analysis (i) based on the zeroth and first weighted moments, and $s = 0.15$ s^{-1} for the analysis (ii) using the zeroth, first and second weighted moments, the parameter values are determined as:

(i) from Eqs. (1.17) and (1.18)

$$\bar{\tau} = 6.6 \text{ s} \quad \text{and} \quad N_D = 0.020$$

(ii) from Eqs. (1.18) and (1.19)

$$\bar{\tau} = 6.7 \text{ s} \quad \text{and} \quad N_D = 0.024.$$

Transfer function fitting

With the measured signals, the transfer function, $F(s)$, is evaluated from Eq. (1.2). Then, as suggested by Eq. (1.25), $F(s)$ is plotted as $-[\ln F(s)]^{-1}$ versus $s[\ln F(s)]^{-2}$ in Figure 1.6. If we examine the graph closely, we will find that the points with large s values are crowded together toward the origin of the graph and the so-called straight line actually crosses the y-axis at a very small negative value. The value of the intercept is so small that the determination of N_D is seriously affected by small errors in the points near the origin or by a slight change in the slope of the straight line.

If Eq. (1.25) is rewritten as:

$$-\frac{\ln F(s)}{s} = -N_D \frac{[\ln F(s)]^2}{s} + \bar{\tau} \tag{1.51}$$

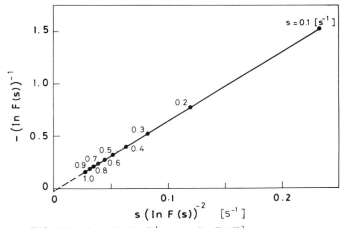

FIGURE 1.6 $-[\ln F(s)]^{-1}$ versus $s[\ln F(s)]^{-2}$ for Example 1.1.

the data, replotted as $-[\ln F(s)]/s$ versus $[\ln F(s)]^2/s$ in Figure 1.7, show that a straight line cannot be drawn over the entire region of s from 0.01 to 1 s^{-1}. But a relatively good straight line exists in the range $s = 0.2$ to 0.4 s^{-1}, from which the parameters are determined as: $\bar{\tau} = 6.7$ s and $N_D = 0.024$.

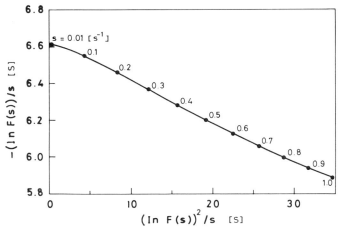

FIGURE 1.7 $-[\ln F(s)]/s$ versus $[\ln F(s)]^2/s$ for Example 1.1.

Figure 1.8 also indicates that a plot based on Eq. (1.26), $[F'(s)/F(s)]^{-2}$ versus s, again does not give a straight line over the range $s = 0$ to $1\,s^{-1}$. But, a linear portion in the range $s = 0.1$ to $0.3\,s^{-1}$ yields $\bar{\tau} = 6.7\,s$ and $N_D = 0.025$.

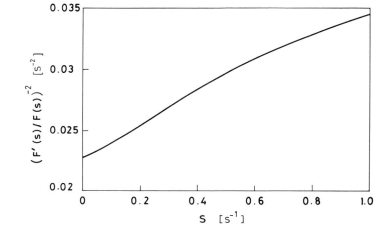

FIGURE 1.8 $[F'(s)/F(s)]^{-2}$ versus s for Example 1.1.

Fourier analysis
The amplitude, A_ω^I, and phase, ϕ_ω^I, for the harmonic component of the input signal are

$$A_\omega^I = [(a_\omega^I)^2 + (b_\omega^I)^2]^{1/2} \Bigg\}$$

and

$$\phi_\omega^I = -\tan^{-1}(b_\omega^I/a_\omega^I)$$

(1.52)

respectively. Similarly, those of the response signal are

$$A_\omega^{II} = [(a_\omega^{II})^2 + (b_\omega^{II})^2]^{1/2} \Bigg\}$$

and

$$\phi_\omega^{II} = -\tan^{-1}(b_\omega^{II}/a_\omega^{II}).$$

(1.53)

The amplitude ratio A_ω and the phase shift ϕ_ω between the two signals are then

$$A_\omega = \frac{A_\omega^{\mathrm{II}}}{A_\omega^{\mathrm{I}}} \Bigg\}$$

and

$$\phi_\omega = \phi_\omega^{\mathrm{II}} - \phi_\omega^{\mathrm{I}}. \Bigg\}$$

(1.54)

The amplitudes, A_ω^{I} and A_ω^{II}, and the ratio, A_ω, are plotted versus ω in Figure 1.9, and the phases, ϕ_ω^{I} and $\phi_\omega^{\mathrm{II}}$, and the phase shift, ϕ_ω, are plotted versus ω in Figure 1.10. (Note that $-\phi_\omega$ is often called a phase lag.)

Fourier analysis should be made in a frequency range where the amplitudes A_ω^{I} and A_ω^{II} are not very small, and the amplitude ratio, A_ω, is appreciably away from unity (or the phase shift ϕ_ω is away from zero). Therefore, both low and high frequencies are found inadequate for Fourier analysis. The parameter values obtained at various frequencies are listed in Table 1.3.

Figure 1.11 is a sensitivity test of the parameter values obtained. The y-axis is the difference Δ, defined by

$$\Delta = \left| \int_0^\infty C_{\mathrm{calc}}^{\mathrm{II}} \exp\left(-i\omega t\right) \mathrm{d}t - \int_0^\infty C_{\mathrm{calc}}^{\mathrm{II}'} \exp\left(-i\omega t\right) \mathrm{d}t \right| \qquad (1.55)$$

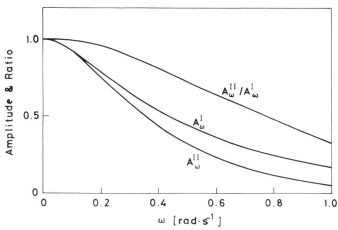

FIGURE 1.9   Amplitudes and amplitude ratio versus ω for Example 1.1.

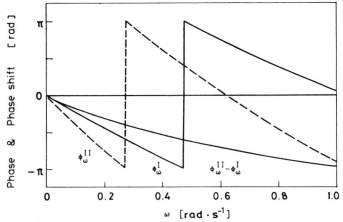

FIGURE 1.10   Phases and phase shift versus ω for Example 1.1.

TABLE 1.3
The values of $\bar{\tau}$ and N_D obtained from Fourier analysis.

ω (rad s^{-1})	$\bar{\tau}$ (s)	N_D
0.1	6.7	0.013
0.2	6.7	0.021
0.3	6.7	0.028
0.4	6.7	0.029
0.5	6.7	0.030
0.6	6.6	0.029
0.7	6.6	0.029
0.8	6.6	0.030

where C_{calc}^{II} is a response signal predicted using the values of $\bar{\tau}$ and N_D listed in Table 1.3, and $C_{calc}^{II'}$ is that calculated using either $1.1\bar{\tau}$ and N_D, or $\bar{\tau}$ and $1.1N_D$ values. Figure 1.11 demonstrates explicitly that the difference, Δ, is much larger for a 10% increase in $\bar{\tau}$ values than the corresponding increase in N_D values. This indicates that $\bar{\tau}$ has a greater effect upon the shape of the response signal than N_D, and consequently, $\bar{\tau}$ may be determined more accurately than N_D.

The $\bar{\tau}$ curve reaches a maximum value at $\omega = 0.32$ rad s^{-1}, and this indicates that $\bar{\tau}$ is determined most accurately at this ω value, or in the

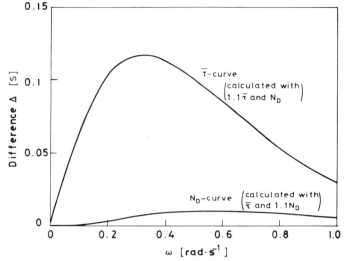

FIGURE 1.11   Sensitivity test in $\bar{\tau}$ and N_D values obtained for Example 1.1.

range, for example, $\omega = 0.3$ to 0.35 rad s^{-1}. Similarly, N_D should be estimated in the range $\omega = 0.5$ to 0.7 rad s^{-1} from the N_D curve. The parameter values are then found to be $\bar{\tau} = 6.7$ s and $N_D = 0.029$.

Curve fitting in the time domain
Curve fitting between the response signal measured and that predicted is
first tested over the central time intervals. Table 1.4 lists the parameter
values determined from the curve fitting and the root-mean-square-errors
between the two signal curves, as defined according to Eq. (1.56).

$$\epsilon = \left[\frac{\displaystyle\int_{t_1}^{t_2} (C_{\text{expt}}^{\text{II}} - C_{\text{calc}}^{\text{II}})^2 \, dt}{\displaystyle\int_{t_1}^{t_2} (C_{\text{expt}}^{\text{II}})^2 \, dt} \right]^{1/2} . \tag{1.56}$$

Figure 1.12 is a plot of N_D versus $\bar{\tau}$ on a map of the root-mean-square-

TABLE 1.4

The values of $\bar{\tau}$ and N_D obtained from curve-fitting and one-point-fitting in the time domain.

	Time interval		$\bar{\tau}$ (s)	N_D	ϵ
	From t_1 (s)	to t_2 (s)			
Curve-fitting	7.5	14.0	6.7	0.030	0.02
	9.0	11.0	6.7	0.032	0.01
	6.5	19.0	6.7	0.031	0.02
	5.0	33.0	6.7	0.030	0.02
One-point-fitting at	t (s)				
	7.0		6.6	0.030	0.02
	10.0[a]		6.9	0.030	0.01

[a] Time when $C_{\mathrm{expt}}^{\mathrm{II}}(t)$ is at the highest peak.

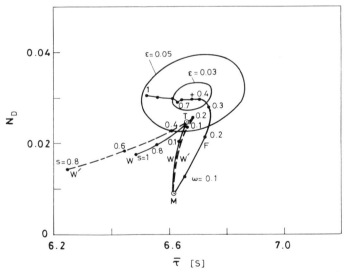

FIGURE 1.12   Error map in the plot of N_D versus $\bar{\tau}$ obtained from curve-fitting in the time domain, for Example 1.1 (least-error corresponds to point labeled +); line W shows the values obtained by the weighted moment method (using zeroth and first moments); line W' shows the weighted moment method (zeroth, first and second moments); line F shows the Fourier analysis; point M shows the moment method; and point T shows transfer function fitting.

error, when the two curves C_{expt}^{II} and C_{calc}^{II} are compared over the time interval $t_1 = 5$ to $t_2 = 33$ s. From the least error point (labeled +), the parameters are found to be $\bar{\tau} = 6.7$ s and $N_D = 0.030$.

Parameter estimation by curve fitting can also be made at any arbitrary single time within the domain. This is tested at $t = 10$ s when the response curve measured reaches a maximum, and at $t = 7$ s in the frontal portion of the response signal. In both cases, the parameter values are, as listed in Table 1.4, in good agreement with those obtained from curve fitting over the central time intervals.

Figure 1.12 compares the values of $\bar{\tau}$ and N_D estimated using the moment method, weighted moment method, transfer function fitting, Fourier analysis and real-time methods of analysis. As the contour map reveals, the parameter values determined using the time domain analysis is, by far, the most accurate. As depicted, the parameter values predicted by Fourier analysis are close to those from the time domain analysis, while the moment method yields less accurate values. On the other hand, the transfer function fitting and weighted moment methods are found, even with their proper values of s, to give data which are not as good as the Fourier analysis.

In Figure 1.13, the response signal measured is compared with those predicted using parameter values obtained by the different methods of analysis. As shown, Curve A (moment method) deviates significantly from Curve E (time domain) which matches well with the measured response signal. Curve D (Fourier analysis) and Curve E almost overlap with each other, while Curve B (weighted moment) and Curve C (transfer function fitting) are between Curve A and Curves D and E.

(End of Example)

1.2 Adsorption Chromatography

When mass transfer takes place from the fluid surrounding a particle into the particle or vice versa, it has usually been assumed that the intraparticle concentration of the mass transferring species is concentric (the same as radial symmetry or center symmetry).

According to the Dispersion-Concentric model (D–C model: fluid in dispersed plug flow and concentric intraparticle concentration), physical adsorption in a packed bed with an imposed signal is described by the

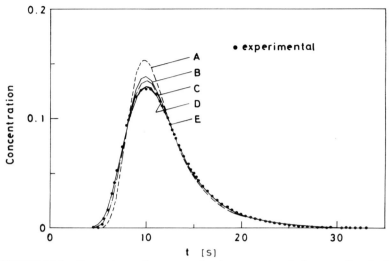

FIGURE 1.13   Comparison of response signal measured and those predicted with the following $\bar{\tau}$ and N_D values, for Example 1.1:

Curve	$\bar{\tau}$ (s)	N_D	ϵ	Method of analysis
A	6.62	0.0087	0.156	Moment
B	6.64	0.020	0.063	Weighted moment (from 0th and 1st)
C	6.66	0.024	0.042	Transfer function
D	6.73	0.029	0.029	Fourier analysis
E	6.68	0.030	0.025	Time domain

following equations:

$$\frac{\partial C}{\partial t} = D_{ax} \frac{\partial^2 C}{\partial x^2} - U \frac{\partial C}{\partial x} - \frac{a}{\epsilon_b} D_e \left(\frac{\partial c}{\partial r}\right)_R \tag{1.57}$$

$$\epsilon_p \frac{\partial c}{\partial t} = D_e \frac{1}{r^2} \frac{\partial}{\partial r}\left(r^2 \frac{\partial c}{\partial r}\right) - \rho_p \frac{\partial c_{ad}}{\partial t} \tag{1.58}$$

and

$$D_e \frac{\partial c}{\partial r} = k_f(C - c) \qquad \text{at} \quad r = R \tag{1.59}$$

with

$$C = c = c_{ad} = 0 \qquad \text{at} \quad t = 0$$

$$\frac{\partial c}{\partial r} = 0 \qquad \text{at} \quad r = 0$$

where

a = particle surface area per unit volume of packed bed;
   $a = 3(1 - \epsilon_b)/R$ for spherical particles
C = tracer concentration in the bulk fluid
c = tracer concentration in the intraparticle pore volume
c_{ad} = amount adsorbed in the particle
D_{ax} = axial dispersion coefficient of the adsorbing species
D_e = intraparticle effective diffusivity
k_f = particle-to-fluid mass transfer coefficient
R = particle radius
r = radial distance variable
U = interstitial fluid velocity
ϵ_b = bed void fraction
ϵ_p = intraparticle void fraction
ρ_p = particle density.

When c is small, the physical adsorption rate is assumed to be first-order

$$\frac{\partial c_{ad}}{\partial t} = k_a \left(c - \frac{c_{ad}}{K_A} \right) \tag{1.60}$$

where

K_A = adsorption equilibrium constant
k_a = adsorption rate constant.

Under the assumption of an infinite bed, the solution in the Laplace domain is

$$F(s) = \exp\left[\frac{1}{2}\left(\frac{LU}{D_{ax}} - \hat{\sigma}_B \right) \right] \tag{1.61}$$

where

$$\hat{\sigma}_{B} = \frac{LU}{D_{ax}} \left[1 + \frac{4D_{ax}}{U^2} (s + q) \right]^{1/2} \tag{1.61a}$$

$$q = \frac{D_e a}{\epsilon_b R} \frac{1}{\dfrac{D_e}{k_f R} + \dfrac{1}{\phi_a \coth \phi_a - 1}} \tag{1.61b}$$

$$\phi_a = R \left[\frac{s}{D_e} \left(\epsilon_p + \frac{\rho_p k_a K_A}{K_A s + k_a} \right) \right]^{1/2}. \tag{1.61c}$$

Also, note that $F(in\pi/\tau)$, which is needed for the prediction of the response signal, is expressed as:

$$F \left(i \frac{n\pi}{\tau} \right) = R_n + iI_n \tag{1.62}$$

where

$$R_n = \exp \left[\left(\frac{U}{2D_{ax}} - \theta\eta \right) L \right] \cos (\theta L) \tag{1.62a}$$

$$I_n = -\exp \left[\left(\frac{U}{2D_{ax}} - \theta\eta \right) L \right] \sin (\theta L) \tag{1.62b}$$

$$\theta = \left\{ \frac{\alpha}{2} [(1 + \gamma^2)^{1/2} - 1] \right\}^{1/2} \tag{1.62c}$$

$$\eta = \left[\frac{(1 + \gamma^2)^{1/2} + 1}{(1 + \gamma^2)^{1/2} - 1} \right]^{1/2} \tag{1.62d}$$

$$\gamma = \frac{\beta}{\alpha} \tag{1.62e}$$

$$\alpha = \left(\frac{U}{2D_{ax}} \right)^2 + \frac{k_f a}{\epsilon_b D_{ax}} \left(1 - \frac{X \cos M \sinh N + Y \sin M \cosh N}{X^2 + Y^2} \right) \tag{1.62f}$$

$$\beta = \frac{\omega}{D_{ax}} - \frac{k_f a}{\epsilon_b D_{ax}} \left(\frac{X \sin M \cosh N - Y \cos M \sinh N}{X^2 + Y^2} \right) \tag{1.62g}$$

$$X = \frac{D_e}{k_f R} (N \cos M \cosh N - M \sin M \sinh N) + \left(1 - \frac{D_e}{k_f R}\right) \cos M \sinh N \tag{1.62h}$$

$$Y = \frac{D_e}{k_f R} (N \sin M \sinh N + M \cos M \cosh N) + \left(1 - \frac{D_e}{k_f R}\right) \sin M \cosh N \tag{1.62i}$$

$$M = \frac{R}{2^{1/2}} [(v^2 + w^2)^{1/2} - w]^{1/2} \tag{1.62j}$$

$$N = \frac{R}{2^{1/2}} [(v^2 + w^2)^{1/2} + w]^{1/2} \tag{1.62k}$$

$$v = \frac{\omega}{D_e} \left(\epsilon_p + \frac{\rho_p k_a^2 K_A}{k_a^2 + K_A^2 \omega^2} \right) \tag{1.62l}$$

$$w = \frac{\omega^2}{D_e} \frac{\rho_p k_a K_A^2}{K_A^2 \omega^2 + k_a^2} \tag{1.62m}$$

$$\omega = \frac{n\pi}{\tau}. \tag{1.62n}$$

1.2.1 Parameter Estimation from First and Second Central Moments

The five parameters, D_{ax}, D_e, k_f, k_a and K_A, are involved in first-order reversible adsorption. Kubin [8] and Kucera [9] showed that all five parameters were determinable, in principle, from five moments generated from a single measurement. However, in practice, higher moments have magnified errors and it is not feasible to determine so many parameters from a single measurement. This is demonstrated in Example 1.2.

Schneider and Smith [10], instead, proposed a method for estimating the parameter values from the first and second central moments based on the response signals from a series of measurements with varied flow rates and particle sizes.

The difference in first moment between the response and input signals is

$$M_1^{II} - M_1^{I} = \frac{L}{U}(1 + \delta_0)$$ (1.63)

where

$$\delta_0 = \frac{aR}{3\epsilon_b}(\epsilon_p + \rho_p K_A).$$ (1.63a)

Similarly, the difference in second central moment or variance, σ_M^2, is

$$\sigma_M^2 = \mu_2^{II} - \mu_2^{I}$$

$$= \frac{2L}{U}\left[\delta_1 + D_{ax}\frac{(1 + \delta_0)^2}{U^2}\right]$$ (1.64)

where

$$\delta_1 = \frac{aR}{3\epsilon_b}\left[\frac{\rho_p(K_A)^2}{k_a} + (\epsilon_p + \rho_p K_A)^2\left(\frac{1}{5D_e} + \frac{1}{k_f R}\right)\frac{R^2}{3}\right].$$ (1.64a)

The method of Schneider and Smith [10] begins by plotting $M_1^{II} - M_1^{I}$ versus L/U and $\sigma_M^2/(2L/U)$ versus U^{-2} for a series of measurements at various flow rates. The K_A value is estimated from the slope of the straight line passing through the origin of the graph of the first moment. A straight line is also drawn for the plot of the variances, and then δ_1 and D_{ax} can be obtained from the intercept and slope of the straight line, respectively. The analysis is based on the assumption that both D_{ax} and k_f are not influenced by U. This assumption is satisfied if the measurements are made at low flow rates.

The values of δ_1 obtained for various particle sizes are plotted against R^2. If $k_f R$ is constant, independent of R, a straight line is drawn. If the value of $k_f R$ is known, the values of k_a and D_e can be determined from the intercept and slope of the straight line. The assumption of a constant $k_f R$ value is, as seen from Eq. (4.11), valid for measurements made at low Reynolds numbers. The method of analysis is demonstrated in Example 1.2.

1.2.2 Parameter Estimation by Curve Fitting in the Time Domain

The techniques outlined in Section 1.2.1 are far more superior than the moment method proposed by Kubin [8] and Kucera [9], which employs higher moments in the estimation of adsorption parameters. As demonstrated in Example 1.1, however, the method of analysis in the time domain is, among all the analytical techniques, the most reliable. The method will be employed, therefore, for parameter estimation in adsorption chromatography. The response signal may be predicted, using the measured input signal, $C_{\text{expt}}^{\text{I}}(t)$, and $F(in\pi/\tau)$ given by Eq. (1.62), from either Eq. (1.31) or Eq. (1.42). The predicted signal, $C_{\text{calc}}^{\text{II}}(t)$, is then compared with the measured signal, $C_{\text{expt}}^{\text{II}}(t)$, in the time domain.

Of the five parameters, D_{ax}, D_{e}, k_{f}, k_{a} and K_{A}, mentioned in Section 1.2.1, k_{f} can be estimated using Eq. (4.11), k_{a} for physical adsorption is usually very large [2, 10] and may be assumed to equal infinity [11], thus, the parameters remaining to be determined are D_{ax}, D_{e} and K_{A}.

Note that when $k_{\text{a}} = \infty$, Eqs. (1.58) and (1.60) become

$$(\epsilon_{\text{p}} + \rho_{\text{p}} K_{\text{A}})\frac{\partial c}{\partial t} = D_{\text{e}}\frac{1}{r^2}\frac{\partial}{\partial r}\left(r^2\frac{\partial c}{\partial r}\right) \tag{1.65}$$

and consequently Eq. (1.61c) reduces to

$$\phi_{\text{a}} = R\left[\frac{s}{D_{\text{e}}}(\epsilon_{\text{p}} + \rho_{\text{p}} K_{\text{A}})\right]^{1/2}. \tag{1.65a}$$

Also, both M and N defined by Eqs. (1.62j) and (1.62k) become

$$M = N = R\left[\frac{\omega}{2D_{\text{e}}}(\epsilon_{\text{p}} + \rho_{\text{p}} K_{\text{A}})\right]^{1/2}. \tag{1.65b}$$

Based on the above equations, Wakao et al. [12] have shown that the value of K_{A} and a relationship between D_{ax} and D_{e} are the only accurate results obtainable from the time domain analysis of a single measurement. They have also shown that if measurements are made with various flow rates, values of D_{ax} and D_{e} could be determined from a series of D_{ax}–D_{e} relationships obtained for different flow rates. This is also demonstrated in Example 1.2. (The Fortran programs for the computations of $C_{\text{calc}}^{\text{II}}(t)$ from Eqs. (1.42), (1.45) and (1.61), and ϵ from Eq. (1.48) are listed in Appendix B.)

Example 1.2

Adsorption chromatography measurements were made at 20°C and atmospheric pressure by imposing nitrogen on a stream of carrier gas of hydrogen in a packed bed ($L = 20.4$ cm and $\epsilon_b = 0.38$) of spherical activated carbon particles (0.2 cm in size and $\epsilon_p = 0.59$). Table 1.5 lists the measured input and response signals of the four runs at $Re = 0.051$, 0.11, 0.30 and 0.47. For illustration, the data for Run 3 are plotted in Figure 1.14 in normalized form, as defined by Eq. (1.50). The transfer function is given by Eq. (1.61). Find what parameters are determinable. The molecular diffusion coefficient, D_v, for the nitrogen–hydrogen system at 20°C and atmospheric pressure is $0.76 \times 10^{-4}\,\mathrm{m^2\,s^{-1}}$.

SOLUTION

Moment method
The first moments, M_1^{II} and M_1^{I}, are calculated from the measured input and response signals, respectively. Figure 1.15 is a plot of $M_1^{II}-M_1^{I}$ versus L/U for all four runs. From the straight line drawn in the graph, $\rho_p K_A$ is found to be 5.30.

The second central moments are also evaluated and plotted, in Figure 1.16, as $\sigma_M^2/(2L/U)$ versus U^{-2}. If we assume that the straight line drawn in the graph represents the data points, we obtain, with the aid of a value of $k_f R$ predicted from Eq. (4.11) and the assumption that $k_a = \infty$, that $D_e = 0.53 \times 10^{-6}\,\mathrm{m^2\,s^{-1}}$ and $D_{ax} = 0.46 \times 10^{-4}\,\mathrm{m^2\,s^{-1}}$ or $\epsilon_b D_{ax}/D_v = 0.23$. It should be noted, however, that the four points in Figure 1.16 lie on a somewhat convex curve. Strictly speaking, it is not correct to represent them by a straight line.

Curve fitting in the time domain
Assuming that $k_a = \infty$ and using Eq. (4.11) for k_f, we consider first the effect of D_{ax} on K_A. Since D_e is not yet known, calculations are made based on the assumed values of $D_e = 0.4 \times 10^{-6}$ and $1 \times 10^{-6}\,\mathrm{m^2\,s^{-1}}$. The error map, in Figure 1.17 for Run 3, shows that for either value of D_e, and for any value of D_{ax} in the range $\epsilon_b D_{ax}/D_v = 0$ to 0.6, the least-error contour corresponds to $\rho_p K_A = 5.29$. Figure 1.18 is a similar error map displaying the effect of $\rho_p K_A$ on D_e for $\epsilon_b D_{ax}/D_v = 0.2$ and 0.4. Again the least-error contour corresponds to $\rho_p K_A = 5.29$. Therefore, we conclude that a single run determines an accurate value of adsorption equilibrium constant.

TABLE 1.5

Adsorption gas chromatography data. Input and response readings.

Time $n\Delta t$ (s)	n	Run 1 $Re = 0.051$ $U = 0.702\ \mathrm{cm\,s^{-1}}$		Run 2 $Re = 0.11$ $U = 1.55\ \mathrm{cm\,s^{-1}}$		Run 3 $Re = 0.30$ $U = 4.20\ \mathrm{cm\,s^{-1}}$		Run 4 $Re = 0.47$ $U = 6.54\ \mathrm{cm\,s^{-1}}$	
		Input	Response	Input	Response	Input	Response	Input	Response
0	0	0.0	—	0.0	—	0.0	—	0.0	—
1	1	21.9	—	2.0	—	6.5	—	3.3	—
2	2	91.4	—	35.5	—	30.5	—	41.3	—
3	3	171.9	—	110.0	—	81.5	—	119.3	—
4	4	208.4	—	187.9	—	133.0	—	193.3	—
5	5	306.8	—	249.9	—	184.0	—	239.3	—
6	6	335.3	—	283.9	—	225.5	—	262.8	—
7	7	346.8	—	300.9	—	251.0	—	265.8	—
8	8	345.3	—	297.9	—	270.5	—	239.8	—
9	9	331.3	—	280.9	—	281.0	—	216.8	—
10	10	313.2	—	254.4	—	282.5	—	173.3	—
11	11	289.7	—	227.4	—	273.5	—	130.7	—
12	12	267.2	—	200.3	—	257.0	—	89.2	—
13	13	244.7	—	174.3	—	238.0	—	70.2	—
14	14	220.6	—	151.8	—	207.0	—	47.7	—
15	15	199.1	—	130.3	—	186.0	—	37.7	—
16	16	180.1	—	111.8	—	158.5	—	27.7	—
17	17	160.6	—	94.3	—	138.5	—	22.2	—
18	18	144.5	—	80.8	—	112.0	—	18.2	—
19	19	129.5	—	69.3	—	99.0	0.0	15.2	—
20	20	114.5	—	58.8	—	83.5	0.6	13.2	—
21	21	103.5	—	49.7	—	73.5	0.8	11.7	—
22	22	92.5	—	42.7	—	60.0	1.1	10.2	—
23	23	82.9	0.0	35.7	—	52.5	1.3	9.2	—

TABLE 1.5 (Continued)

Time $n\Delta t$ (s)	Run 1 $Re = 0.051$ $U = 0.702\,\text{cm s}^{-1}$		Run 2 $Re = 0.11$ $U = 1.55\,\text{cm s}^{-1}$		Run 3 $Re = 0.30$ $U = 4.20\,\text{cm s}^{-1}$		Run 4 $Re = 0.47$ $U = 6\ 54\,\text{cm s}^{-1}$	
n	Input	Response	Input	Response	Input	Response	Input	Response
24	74.4	0.5	30.7	—	45.0	1.7	8.7	—
25	66.4	0.4	25.2	—	40.0	2.2	8.2	—
26	59.4	0.9	21.7	—	34.5	3.2	7.7	0.0
27	52.8	0.9	18.2	—	31.5	4.7	7.2	0.5
28	47.3	1.4	15.7	—	27.5	8.3	6.7	1.0
29	42.3	1.5	12.6	0.0	24.0	12.3	6.2	1.0
30	38.3	3.3	11.1	1.1	22.0	17.3	5.7	1.6
31	34.8	4.3	9.1	1.6	19.5	23.3	5.2	2.6
32	30.2	6.2	7.6	2.1	17.5	30.9	4.7	3.6
33	27.7	8.2	6.6	2.6	16.0	40.4	4.7	5.1
34	25.2	10.7	6.1	4.1	14.5	50.4	4.2	7.6
35	22.2	13.2	5.1	5.1	13.0	61.0	4.2	10.1
36	20.1	16.1	4.6	6.2	12.0	72.5	3.7	13.1
37	18.6	21.1	4.0	8.2	11.0	83.0	3.6	17.1
38	17.1	25.6	3.5	11.2	10.5	93.0	3.1	21.2
39	15.6	30.0	3.0	14.2	9.5	100.1	3.1	26.2
40	14.0	36.5	2.5	19.2	9.0	107.1	2.6	31.2
41	12.5	41.0	2.5	24.3	8.0	111.1	2.6	37.7
42	11.5	47.0	2.0	30.8	7.5	114.1	2.1	43.2
43	10.5	53.9	2.0	37.3	7.0	115.2	2.1	49.2
44	9.5	60.9	1.5	45.3	6.5	114.2	2.1	55.2
45	8.4	68.9	1.4	53.8	6.0	111.2	2.1	62.3
46	7.4	75.8	1.4	62.8	5.5	106.7	1.6	67.3
47	7.1	83.3	1.4	71.4	5.5	101.3	1.6	73.3

48	6.9	89.8	1.4	79.9	5.0	94.3	1.6	78.3
49	6.5	95.8	1.1	89.4	5.0	87.8	1.6	83.8
50	6.3	102.7	0.9	97.4	4.5	80.4	1.6	87.3
51	5.8	108.7	0.9	105.9	4.5	72.4	1.6	90.3
52	5.3	115.2	0.9	113.4	4.0	64.4	1.6	92.3
53	5.3	119.7	0.8	120.5	4.0	56.9	1.6	93.9
54	5.2	124.1	0.8	125.5	3.5	50.5	1.1	94.4
55	4.7	129.1	0.8	130.5	3.5	43.0	1.1	94.4
56	4.2	131.6	0.8	134.0	3.0	37.5	1.1	93.9
57	4.2	133.5	0.8	136.0	3.0	32.0	1.1	92.4
58	4.1	136.0	0.8	137.1	2.5	27.1	1.1	89.9
59	3.8	137.0	1.0	137.6	2.5	22.6	1.1	86.9
60	3.6	138.5	1.3	135.1	2.0	19.1	0.6	83.9
61	3.9	139.4	1.2	132.1	2.0	16.2	0.6	79.5
62	4.0	137.9	1.2	128.6	1.5	13.7	0.5	75.0
63	3.7	137.4	0.9	125.6	1.5	11.2	0.5	71.0
64	3.5	136.8	0.7	120.2	1.0	8.7	0.5	67.0
65	3.5	133.3	0.7	114.7	1.0	6.8	0.5	63.0
66	3.5	131.8	0.7	109.2	1.0	5.3	0.5	58.5
67	3.1	129.8	0.7	103.7	1.0	4.3	0.5	54.0
68	2.9	127.2	0.7	97.2	0.5	3.3	0.5	50.6
69	2.9	123.7	0.6	91.8	0.5	2.4	0.5	45.6
70	2.9	120.7	0.6	85.3	0.5	1.4	0.5	42.1
71	2.5	116.6	0.6	80.3	0.5	0.9	0.4	38.6
72	2.3	112.1	0.6	74.3	0.5	0.4	0.2	34.6
73	2.3	109.1	0.6	68.8	0.5	0.0	0.1	31.6
74	2.3	104.1	0.6	62.3	0.5	—	0.0	28.1
75	2.0	100.0	0.6	56.9	0.5	—	—	25.6
76	1.7	95.5	0.6	52.9	0.5	—	—	23.2
77	1.4	90.5	0.4	47.4	0.5	—	—	20.2
78	1.2	86.4	0.3	42.9	0.5	—	—	17.7
79	0.9	82.9	0.2	38.9	0.5	—	—	15.7

TABLE 1.5 (Continued)

Time $n\Delta t$ (s)		Run 1 Re = 0.051 U = 0.702 cm s⁻¹		Run 2 Re = 0.11 U = 1.55 cm s⁻¹		Run 3 Re = 0.30 U = 4.20 cm s⁻¹		Run 4 Re = 0.47 U = 6.54 cm s⁻¹	
n		Input	Response	Input	Response	Input	Response	Input	Response
80		0.6	79.4	0.0	34.9	0.5	—	—	14.2
81		0.6	74.9	—	31.0	0.5	—	—	12.7
82		0.6	69.8	—	28.0	0.5	—	—	11.2
83		0.8	66.8	—	25.0	0.0	—	—	9.8
84		0.5	62.8	—	23.0	—	—	—	8.8
85		0.3	59.7	—	20.5	—	—	—	7.3
86		0.0	57.2	—	18.6	—	—	—	6.3
87		—	52.7	—	16.1	—	—	—	5.3
88		—	49.7	—	14.1	—	—	—	4.8
89		—	46.6	—	13.1	—	—	—	4.3
90		—	43.6	—	11.6	—	—	—	3.8
91		—	40.1	—	10.1	—	—	—	3.4
92		—	37.5	—	9.2	—	—	—	2.9
93		—	35.5	—	7.7	—	—	—	2.4
94		—	33.5	—	6.7	—	—	—	1.9
95		—	30.5	—	5.7	—	—	—	1.4
96		—	28.4	—	5.2	—	—	—	0.9
97		—	26.4	—	4.8	—	—	—	0.9
98		—	24.4	—	4.3	—	—	—	0.4
99		—	22.3	—	3.8	—	—	—	0.5
100		—	20.8	—	3.8	—	—	—	0.5
101		—	19.8	—	3.3	—	—	—	0.0
102		—	18.3	—	2.3	—	—	—	—

	2.5	5.0	1.25	2.5	0.3125	1.25	0.3125	0.625
103	—	16.2	—	1.9	—	—	—	—
104	—	14.2	—	1.9	—	—	—	—
105	—	13.2	—	1.4	—	—	—	—
106	—	12.7	—	1.4	—	—	—	—
107	—	11.1	—	0.9	—	—	—	—
108	—	9.6	—	0.9	—	—	—	—
109	—	8.6	—	1.0	—	—	—	—
110	—	7.0	—	0.5	—	—	—	—
111	—	6.5	—	0.0	—	—	—	—
112	—	5.5	—	—	—	—	—	—
113	—	5.0	—	—	—	—	—	—
114	—	4.9	—	—	—	—	—	—
115	—	4.4	—	—	—	—	—	—
116	—	3.4	—	—	—	—	—	—
117	—	3.3	—	—	—	—	—	—
118	—	3.3	—	—	—	—	—	—
119	—	3.8	—	—	—	—	—	—
120	—	3.3	—	—	—	—	—	—
121	—	2.7	—	—	—	—	—	—
122	—	2.2	—	—	—	—	—	—
123	—	1.7	—	—	—	—	—	—
124	—	2.1	—	—	—	—	—	—
125	—	0.6	—	—	—	—	—	—
126	—	0.6	—	—	—	—	—	—
127	—	0.6	—	—	—	—	—	—
128	—	0.0	—	—	—	—	—	—
Time interval Δt (s)	2.5	5.0	1.25	2.5	0.3125	1.25	0.3125	0.625

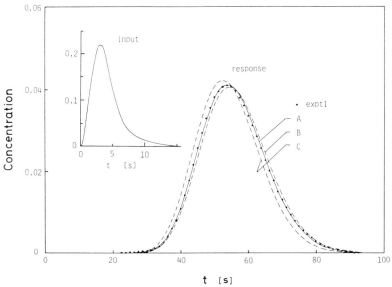

t [s]

FIGURE 1.14   Input and response signals measured from adsorption chromatography; response signals predicted for Run 3 in Example 1.2:

	Response signal predicted with			
Curve	$\rho_p K_A$	$\epsilon_b D_{ax}/D_v$	D_e $(m^2 s^{-1})$	ϵ
A	5.29	0.24	0.63×10^{-6}	0.016
B	5.37	0.24	0.63×10^{-6}	0.05
C	5.11	0.24	0.63×10^{-6}	0.10

Using the value of $\rho_p K_A$ obtained, the error map for various values of $\epsilon_b D_{ax}/D_v$ and D_e is shown in Figure 1.19. For example, the values of these two parameters within the shaded part of the figure show that $C^{II}_{calc}(t)$ differs from $C^{II}_{expt}(t)$ by a root-mean-square-error, ϵ, defined by Eq. (1.47), of less than 0.025. The region with the least error may be visualized as a valley in a three-dimensional error map. The valley expands with an increase in D_e. The graph indicates that simultaneous determination of the two parameters D_e and D_{ax} is not feasible. However, if we know one of them, the value of the other can be determined from the contour for $\epsilon \leqslant 0.025$.

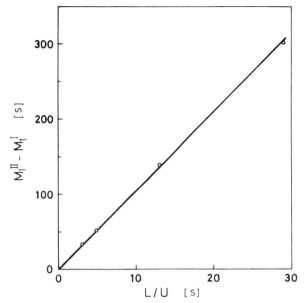

FIGURE 1.15 $M_1^{II} - M_1^{I}$ versus L/U for Example 1.2.

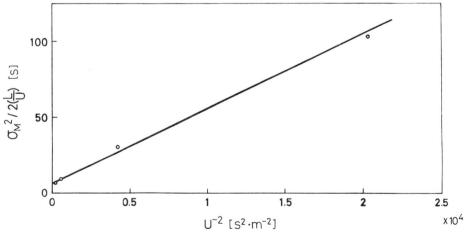

FIGURE 1.16 $\sigma_M^2/(2L/U)$ versus U^{-2} for Example 1.2.

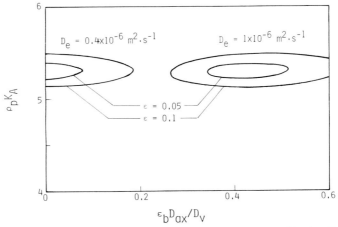

FIGURE 1.17   Error map in the plot of $\rho_p K_A$ versus $\epsilon_b D_{ax}/D_v$ for Run 3 in Example 1.2.

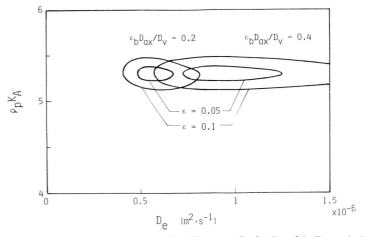

FIGURE 1.18   Error map in the plot of $\rho_p K_A$ versus D_e for Run 3 in Example 1.2.

Figure 1.20 shows the error map for the four runs with varied flow rates. The contours for $\epsilon \leqslant 0.025$ are steep with respect to D_e at high flow rates and nearly flat at low flow rates. Thus, D_e has a larger effect on the response curve at high flow rates, but is not important at low flow rates. In the laminar flow range, D_{ax} is constant, not depending on flow rate, so

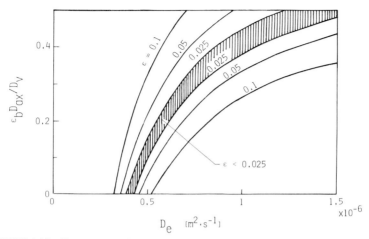

FIGURE 1.19   Error map in the plot of $\epsilon_b D_{ax}/D_v$ versus D_e for Run 3 in Example 1.2.

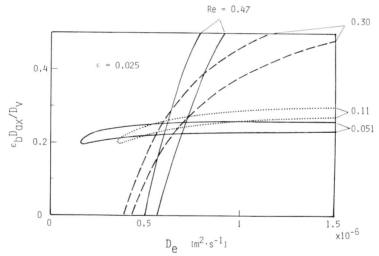

FIGURE 1.20   Error map in the plot of $\epsilon_b D_{ax}/D_v$ versus D_e for various flow rates, for Example 1.2.

that the best values of D_{ax} and D_e correspond to the basin where all four valleys overlap. This is shown more clearly in Figure 1.21, which is a map of the arithmetic mean error for all the runs. The least-error point (+ in the graph) corresponds to $\epsilon_b D_{ax}/D_v = 0.24$ and $D_e = 0.63 \times 10^{-6}$ m^2 s^{-1}.

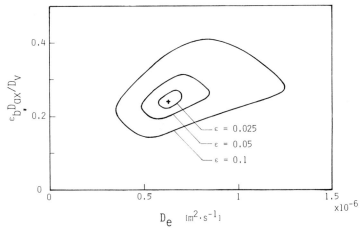

FIGURE 1.21   Map of the mean-error in the plot of $\epsilon_b D_{ax}/D_v$ versus D_e for Example 1.2 (least-error corresponds to point labeled +).

With these parameter values determined, the predicted response curve, $C_{calc}^{II}(t)$, is that labeled A in Figure 1.14. It agrees well with the experimental curve; the root-mean-square-error is $\epsilon = 0.016$. Curves B and C illustrate the large effect of $\rho_p K_A$ on the response. A 1.5% increase in $\rho_p K_A$ from 5.29 to 5.37 increases the error, ϵ, from 0.016 to 0.05. Similarly, a 3.4% drop in this parameter, from 5.29 to 5.11, increases the error to 0.10.

In adsorption chromatography, the adsorption equilibrium constant has the largest effect on the response curve. As illustrated, an accurate value of K_A can be determined from a single measurement. In order to establish accurate values of the axial dispersion coefficient and the effective diffusivity in the particle, input–response curves need to be measured over a range of fluid velocities. These conclusions apply only when the particle-to-fluid mass transfer process and the adsorption rate have little effect on the response curve. Such conditions are usually met for physical adsorption of gas in beds of small adsorbent particles.

It should be noted also that the intraparticle diffusivity cannot be accurately determined solely from response curves measured at low flow rates. This is evident from Figure 1.20 where the least error valley extends horizontally. Error maps such as Figures 1.20 and 1.21 are particularly useful for determining the range of operating conditions best suited for the evaluation of accurate parameter values.

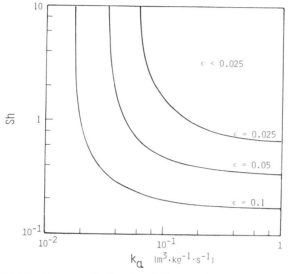

FIGURE 1.22   Error map in the plot of Sh versus k_a for Run 3 in Example 1.2.

The measured response signal may be used to evaluate the influence of the adsorption rate constant. Calculated response curves for Run 3 are obtained using $\epsilon_b D_{ax}/D_v = 0.24$, $D_e = 0.63 \times 10^{-6}$ m^2 s^{-1} and with various assumed values of Sherwood number and k_a. Figure 1.22 is the resulting error map. The curves show that, for $Sh > 2$, for example, any value for k_a greater than about 0.09 m^3 kg^{-1} s^{-1} will have a small $(\epsilon < 0.025)$ effect on the response signal.

(End of Example)

Figure 1.20 shows that the $\epsilon_b D_{ax}/D_v$-D_e contour becomes flat as the flow rate decreases. The contours are regarded as the valleys on a three-dimensional error map. If we take a look at a moment of the system, the moment value must be constant everywhere along the least-error (or deepest valley) line in the middle of the contour. For example, according to Eq. (1.64), the second central moments or variances should be constant along the deepest valley. In other words, the following condition prevails:

$$d(\sigma_M^2) = 0. \tag{1.66}$$

Therefore, the slope of the line in a graph of D_{ax} versus D_e is

$$\frac{dD_{ax}}{dD_e} = \frac{1}{20\epsilon_b aR} \left[\frac{\delta_0}{(1 + \delta_0)} \frac{D_v}{D_e} (Sc)(Re) \right]^2. \tag{1.67}$$

Simultaneous determination of D_{ax} and D_e needs at least two curves crossing each other on a graph of D_{ax} versus D_e, one horizontal or nearly horizontal and the other steep. The former with $dD_{ax}/dD_e \simeq 0$ is obviously obtained at low Reynolds numbers. The curve with high dD_{ax}/dD_e is obtained, on the contrary, at high flow rates. If, however, δ_0 or $(\epsilon_p + \rho_p K_A)$ is small, the slope dD_{ax}/dD_e is low even at high flow rates.

The value of $\rho_p K_A$ is accurately determined from a single chromatography measurement. In most cases, D_e/D_v is less than 0.1. Therefore, if the value of $(\epsilon_p + \rho_p K_A)$ is large, probably larger than about unity, both D_{ax} and D_e may be determined from the time domain analysis of the measurements made in a laminar flow range. If the value of $(\epsilon_p + \rho_p K_A)$ is small, however, intraparticle diffusion makes little contribution to the overall transport in adsorption chromatography.

1.2.3   Relationship Between Overall Gaseous Effective Diffusivity and Macropore and Micropore Diffusivities

When a porous pellet has a bidisperse macropore/micropore structure, the simplest model is, of course, the one in which the bidisperse pores are not distinguished, and the concept of overall effective diffusivity is introduced. The analysis in the preceding sections is based on such a model and is called Model I in this section. Molecular sieves have a typical bidisperse pore structure. In macropores, gaseous diffusion is considered to take place, but the micropores have been assumed to be in: (i) gas phase by Antonson and Dranoff [13], Ruckenstein et al. [14], Hashimoto and Smith [15], and Kawazoe and Takeuchi [16]; (ii) adsorbate phase by Kawazoe and Takeuchi [16], Schneider and Smith [17], Kawazoe et al. [18], and Lee and Ruthven [19]. Let us call them Models II and III, respectively.

	Macropores	Micropores
Model II	Gas phase	Gas phase
Model III	Gas phase	Adsorbate phase

In the models, the spherical pellet is assumed to be composed of an assemblage of fine spherical microporous particles.

The fundamental equations for Models I–III are listed together with the first and second central moments for impulse response in Table 1.6. Besides the notation listed below Eq. (1.59), the following symbols are used:

$$c_a = \text{gaseous concentration in the macropores}$$

$c_{ad}, c'_{ad}, c''_{ad} = $ amount adsorbed per unit mass of pellet, used in Models I, II and III, respectively

$c_i = $ gaseous concentration in microporous particle

$D_a = $ gaseous effective diffusivity in the macropores, based on cross-sectional area of the pellet

$D_i = $ gaseous effective diffusivity in the micropores, based on cross-sectional area of microporous particle

$D_{si} = $ effective diffusivity of adsorbate in the micropores, based on cross-sectional area of microporous particle

$K_A, K'_A, K''_A = $ adsorption equilibrium constants, used in Models I, II and III, respectively

$k_a, k'_a, k''_a = $ adsorption rate constants, used in Models I, II and III, respectively

$n_d, n'_d, n''_d = $ molar fluxes from gas to pellet, used in Models I, II and III, respectively

$r_i = $ radial distance variable in microporous particle

$r_0 = $ radius of microporous particle

$\epsilon_a = $ macropore void fraction of pellet

$\epsilon_i = $ micropore void fraction of pellet, accessible for adsorbing

$\epsilon_{ip} = $ gas species

$\epsilon_p = $ volume fraction of microporous particles in pellet

$$\epsilon_a + \epsilon_i.$$

A comparison of the first moments of the three models indicates that

$$\text{Model I} \qquad \text{Model II} \qquad \text{Model III}$$

$$\epsilon_p + \rho_p K_A \;=\; \epsilon_p + \rho_p K'_A \;=\; \epsilon_a + \rho_p K''_A. \qquad (1.68)$$

Therefore,

$$K_A = K'_A = K''_A - \frac{\epsilon_i}{\rho_p}. \qquad (1.69)$$

TABLE 1.6
Models for gas adsorption in a packed bed.

	Model I	Model II	Model III
		Macro–micropore model	
	Single gaseous diffusion model	Gas diffusion in macropores, Gas diffusion in micropores	Gas diffusion in macropores, Adsorbate diffusion in micropores

Particle

Model I:

$$\epsilon_p \frac{\partial c}{\partial t} = \frac{D_e}{r^2}\frac{\partial}{\partial r}\left(r^2\frac{\partial c}{\partial r}\right) - \rho_p \frac{\partial c_{ad}}{\partial t}$$

Model II:

Macropores:
$$\epsilon_a \frac{\partial c_a}{\partial t} = \frac{D_a}{r^2}\frac{\partial}{\partial r}\left(r^2\frac{\partial c_a}{\partial r}\right) - \frac{3\epsilon_{ip}D_i}{r_0}\left(\frac{\partial c_i}{\partial r_i}\right)_{r_0}$$

Microporous particle:
$$\epsilon_i \frac{\partial c_i}{\partial t} = \frac{\epsilon_{ip}D_i}{r_i^2}\frac{\partial}{\partial r_i}\left(r_i^2\frac{\partial c_i}{\partial r_i}\right) - \rho_p \frac{\partial c'_{ad}}{\partial t}$$
$$c_a = (c_i)_{r_0}$$

Model III:

Macropores:
$$\epsilon_a \frac{\partial c_a}{\partial t} = \frac{D_a}{r^2}\frac{\partial}{\partial r}\left(r^2\frac{\partial c_a}{\partial r}\right) - \frac{3\rho_p D_{Si}}{r_0}\left(\frac{\partial c''_{ad}}{\partial r_i}\right)_{r_0}$$

Adsorbate diffusion in micropores:
$$\frac{\partial c''_{ad}}{\partial t} = \frac{D_{Si}}{r_i^2}\frac{\partial}{\partial r_i}\left(r_i^2\frac{\partial c''_{ad}}{\partial r_i}\right)$$

Gas

Model I:
$$\frac{\partial C}{\partial t} = D_{ax}\frac{\partial^2 C}{\partial x^2} - U\frac{\partial C}{\partial x} - \frac{a}{\epsilon_b} n_d$$
$$n_d = D_e\left(\frac{\partial c}{\partial r}\right)_R = k_f[C-(c)_R]$$

Model II:
$$\frac{\partial C}{\partial t} = D_{ax}\frac{\partial^2 C}{\partial x^2} - U\frac{\partial C}{\partial x} - \frac{a}{\epsilon_b} n'_d$$
$$n'_d = D_a\left(\frac{\partial c_a}{\partial r}\right)_R = k_f[C-(c_a)_R]$$

Model III:
$$\frac{\partial C}{\partial t} = D_{ax}\frac{\partial^2 C}{\partial x^2} - U\frac{\partial C}{\partial x} - \frac{a}{\epsilon_b} n''_d$$
$$n''_d = D_a\left(\frac{\partial c_a}{\partial r}\right)_R = k_f[C-(c_a)_R]$$

Adsorption rate	$\dfrac{\partial c_{ad}}{\partial t} = k_a \left(c - \dfrac{c_{ad}}{K_A} \right)$	$\dfrac{\partial c'_{ad}}{\partial t} = k'_a \left(c'_i - \dfrac{c'_{ad}}{K'_A} \right)$	$\dfrac{\rho_P D_{si}}{\epsilon_{ip}} \left(\dfrac{\partial c''_{ad}}{\partial r_i} \right)_{r_o} = k''_a \left[c_a - \dfrac{(c''_{ad})_{r_o}}{K''_A} \right]$
First moment of impulse response	$M_1^{II} = \dfrac{L}{U} \left[1 + \dfrac{aR}{3\epsilon_b}(\epsilon_p + \rho_p K_A) \right]$	$\dfrac{L}{U} \left[1 + \dfrac{aR}{3\epsilon_b}(\epsilon_p + \rho_p K'_A) \right]$	$\dfrac{L}{U} \left[1 + \dfrac{aR}{3\epsilon_b}(\epsilon_a + \rho_p K''_A) \right]$
Second central moment of impulse response	$\mu_2^{II} = \dfrac{2L}{U} \left[\delta_1 + D_{ax} \dfrac{(1+\delta_0)^2}{U^2} \right]$	$\dfrac{2L}{U} \left[\delta'_1 + D_{ax} \dfrac{(1+\delta'_0)^2}{U^2} \right]$	$\dfrac{2L}{U} \left[\delta''_1 + D_{ax} \dfrac{(1+\delta''_0)^2}{U^2} \right]$
where	$\delta_0 = \dfrac{aR}{3\epsilon_b}(\epsilon_p + \rho_p K_A)$	where $\delta'_0 = \dfrac{aR}{3\epsilon_b}(\epsilon_p + \rho_p K'_A)$	where $\delta''_0 = \dfrac{aR}{3\epsilon_b}(\epsilon_a + \rho_p K''_A)$
	$\delta_1 = \dfrac{aR}{3\epsilon_b}\left[\dfrac{\rho_p(K_A)^2}{k_a} + (\epsilon_p + \rho_p K_A)^2 \right]$ $\times \left(\dfrac{1}{5D_e} + \dfrac{1}{k_f R} \right) \dfrac{R^2}{3}$	$\delta'_1 = \dfrac{aR}{3\epsilon_b}\left[\dfrac{\rho_p(K'_A)^2}{k'_a} + (\epsilon_p + \rho_p K'_A)^2 \right]$ $\times \left(\dfrac{1}{5D_a} + \dfrac{1}{k_f R} \right) \dfrac{R^2}{3}$ $+ \dfrac{r_0^2}{15\epsilon_{ip} D_i}(\epsilon_l + \rho_p K'_A)^2$	$\delta''_1 = \dfrac{aR}{3\epsilon_b}\left[\dfrac{r_0(\rho_p K''_A)^2}{3\epsilon_{ip} k''_a} + (\epsilon_a + \rho_p K''_A)^2 \right]$ $\times \left(\dfrac{1}{5D_a} + \dfrac{1}{k_f R} \right) \dfrac{R^2}{3} + \dfrac{r_0^2 \rho_p K''_A}{15 D_{si}}$

Similarly, from a comparison of adsorption terms of the second central moments, the following expression is obtained:

$$
\underset{Model\ I}{\frac{\rho_p(K_A)^2}{k_a}} = \underset{Model\ II}{\frac{\rho_p(K_A')^2}{k_a'}} = \underset{Model\ III}{\frac{r_0(\rho_p K_A'')^2}{3\epsilon_{ip}k_a''}}. \tag{1.70}
$$

Consequently,

$$
k_a = k_a' = \frac{3\epsilon_{ip}k_a''}{r_0\rho_p}\left(1 - \frac{\epsilon_i}{\rho_p K_A''}\right)^2. \tag{1.71}
$$

The effective diffusivity terms of the models are related through Eq. (1.72).

$$
\underset{Model\ I}{\frac{1}{D_e}} = \underset{Model\ II}{\frac{1}{D_a} + \frac{1}{\epsilon_{ip}D_i}\left(\frac{r_0}{R}\right)^2\left(\frac{\epsilon_i + \rho_p K_A'}{\epsilon_p + \rho_p K_A'}\right)^2}
$$

$$
\underset{Model\ III}{= \frac{1}{D_a} + \frac{1}{D_{si}}\left(\frac{r_0}{R}\right)^2\frac{\rho_p K_A''}{(\epsilon_a + \rho_p K_A'')^2}}. \tag{1.72}
$$

Equations (1.69), (1.71) and (1.72) show the relationships between the parameters of the different models. Equation (1.68) also indicates that, as far as the first moments of the models are concerned, the controlling parameters are intraparticle void fraction for gaseous diffusion plus the adsorption equilibrium constant. It is clear from Eq. (1.72) that the D_e value of Model I is close to the D_a value of Model II (because of small r_0/R) and also between the D_a and D_{si} values of Model III (because D_{si} is also small). The parameters used in the models are reciprocally convertible, but unless chromatography measurements are made with varied temperatures, no answer is given to the question about whether the micropores are in the gas phase (Model II) or in the adsorbate phase (Model III).

1.2.4   Assumption of Concentric Intraparticle Concentration

Take a look at an adsorbent particle in a packed bed. When an adsorbing species, imposed as a step-function or a one-shot input on a carrier stream, passes over the particle, the intraparticle concentration of the adsorbing species cannot have radial symmetry. If the tracer species starts to pass over the particle from left to right, for example, it is apparent that more of the species penetrate into the particle from the left hand side.

However, Eq. (1.58) is based on the assumption that the intraparticle concentration is radially symmetric. The purpose of this section is to examine whether the assumption causes any errors in the parameter estimation from adsorption chromatography under isothermal conditions.

The intraparticle concentration of the adsorbing species is expressed in the spherical coordinates shown in Figure 1.23. Writing the gaseous con-centration and adsorbate concentrations as c^* and c_{ad}^*, respectively, the mass balance equation is

$$\epsilon_p \frac{\partial c^*}{\partial t} = D_e \nabla^2 c^* - \rho_p \frac{\partial c_{ad}^*}{\partial t} \qquad (1.73)$$

where

$$\nabla^2 = \frac{1}{r^2} \frac{\partial}{\partial r} \left(r^2 \frac{\partial}{\partial r} \right) + \frac{1}{r^2} \left[\frac{1}{\sin\theta} \frac{\partial}{\partial\theta} \left(\sin\theta \frac{\partial}{\partial\theta} \right) + \frac{1}{\sin^2\theta} \frac{\partial^2}{\partial\Phi^2} \right] \cdot \quad (1.73a)$$

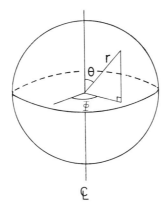

FIGURE 1.23   Spherical coordinates.

When the adsorption is first-order and reversible, the adsorption rate is

$$\frac{\partial c_{ad}^*}{\partial t} = k_a \left(c^* - \frac{c_{ad}^*}{K_A} \right). \tag{1.74}$$

Integrating Eq. (1.73) over a spherical surface with radius r and dividing this by $4\pi r^2$, it follows that:

$$\frac{1}{4\pi r^2} \int_0^{2\pi} d\Phi \int_{-1}^{1} \left(\epsilon_p \frac{\partial c^*}{\partial t} \right) r^2 \, d\cos\theta$$

$$= \frac{1}{4\pi r^2} \int_0^{2\pi} d\Phi \int_{-1}^{1} \left(D_e \nabla^2 c^* - \rho_p \frac{\partial c_{ad}^*}{\partial t} \right) r^2 \, d\cos\theta. \tag{1.75}$$

If the order of differentiation and integration is reversed, and considering $(\partial c^*/\partial\Phi)_{\Phi=0} = (\partial c^*/\partial\Phi)_{\Phi=2\pi}$, Eq. (1.75) becomes

$$\epsilon_p \frac{\partial X}{\partial t} = D_e \frac{1}{r^2} \frac{\partial}{\partial r} \left(r^2 \frac{\partial X}{\partial r} \right) - \rho_p \frac{\partial Y}{\partial t} \tag{1.76}$$

where

$$X = \frac{1}{4\pi} \int_0^{2\pi} d\Phi \int_{-1}^{1} c^* \, d\cos\theta \tag{1.76a}$$

and

$$Y = \frac{1}{4\pi} \int_0^{2\pi} d\Phi \int_{-1}^{1} c_{ad}^* \, d\cos\theta. \tag{1.76b}$$

It is clear that X and Y are the average concentrations of c^* and c_{ad}^*, respectively, over a spherical area $4\pi r^2$ inside the spherical pellet. Equation

(1.76) is identical to Eq. (1.58) which is derived under the assumption that the intraparticle concentration has radial symmetry. Similarly, integrating Eq. (1.74) over a spherical area with radius r and dividing by $4\pi r^2$, we obtain Eq. (1.60).

Therefore, it is concluded that Eqs. (1.57)-(1.60), derived assuming that the concentration has radial symmetry, are valid as long as the adsorption rate is first-order.

1.3   Effect of Dead Volume Associated with Signal Detecting Elements

In chromatography measurements, the input and response signals are measured using detecting elements inserted into the packed bed. The elements are usually placed in a shallow empty section installed in the column. Kaguei et al. [20] examined the effect of dead volume upon parameter estimation.

1.3.1   Packed Beds of Glass Beads

Suppose the detecting elements are inserted into the middle of the dead volume between the packed beds of glass beads, as shown in Figure 1.24. The material balance equations for a tracer injected into the column give:

In the dead volumes (concentration C'), Section 1 ($0 < x < L_D$) and Section 3 ($L + L_D < x < L + 3L_D$)

$$\frac{\partial C'}{\partial t} = D' \frac{\partial^2 C'}{\partial x^2} - u' \frac{\partial C'}{\partial x} \tag{1.77}$$

where D' is the dispersion coefficient and u' is the fluid velocity in the dead volume section.

In the packed beds (concentration C), Section 2 ($L_D < x < L + L_D$) and Section 4 ($x > L + 3L_D$)

$$\frac{\partial C}{\partial t} = D_{ax} \frac{\partial^2 C}{\partial x^2} - U \frac{\partial C}{\partial x} \tag{1.78}$$

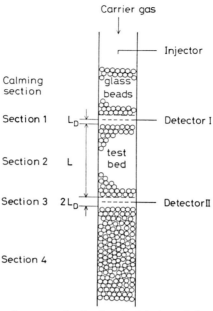

FIGURE 1.24   A column consisted of packed beds and dead volumes associated
with concentration detecting elements.

with

$$C' = C = 0 \qquad \text{at} \quad t = 0$$

$$C = 0 \qquad \text{at} \quad x = \infty.$$

Conditions at the boundary between the packed bed and dead volume
section are also needed. The mass balance equation at the boundary
gives

$$u'C' - D'\frac{\partial C'}{\partial x} = \epsilon_b \left(UC - D_{ax}\frac{\partial C}{\partial x} \right). \qquad (1.79)$$

In the column shown in Figure 1.24, there cannot be a concentration
discontinuity at the boundary. Obviously, $u' = \epsilon_b U$, and consequently, the
dispersion fluxes are also equal at the boundary.

$$C' = C$$

$$D' \frac{\partial C'}{\partial x} = \epsilon_b D_{ax} \frac{\partial C}{\partial x} \Bigg\} \quad \text{at } x = L_D, L + L_D \text{ and } L + 3L_D.$$

If we simply write $C'(x = 0) = C^I$ and $C'(x = L + 2L_D) = C^{II}$, the transfer function is

$$F(s) = \frac{\displaystyle\int_0^\infty C^{II} \exp(-st)\, dt}{\displaystyle\int_0^\infty C^I \exp(-st)\, dt}$$

$$= \frac{\alpha(L_D/L)\, \sigma_D \sigma_B [\delta + \gamma \exp(-\sigma_D)] \exp(\lambda_B + 2\lambda_D)}{\{[\delta + \gamma \exp(-\sigma_D)]^2 [\delta - \gamma \exp(-\sigma_D)] \\ \quad - \delta\gamma[1 - \exp(-2\sigma_D)][\gamma + \delta \exp(-\sigma_D)] \exp(-\sigma_B)\}}$$

$$(1.80)$$

where

$$\alpha = \frac{\epsilon_b D_{ax}}{D'} \tag{1.80a}$$

$$\gamma = \frac{1}{2}\left(\sigma_D - \alpha\sigma_B \frac{L_D}{L}\right) \tag{1.80b}$$

$$\delta = \frac{1}{2}\left(\sigma_D + \alpha\sigma_B \frac{L_D}{L}\right) \tag{1.80c}$$

$$\lambda_B = \frac{1}{2}\left(\frac{LU}{D_{ax}} - \sigma_B\right) \tag{1.80d}$$

$$\lambda_D = \frac{1}{2}\left(\frac{L_D u'}{D'} - \sigma_D\right) \tag{1.80e}$$

$$\sigma_B = \frac{LU}{D_{ax}}\left(1 + \frac{4D_{ax}}{U^2}s\right)^{1/2}$$ (1.80f)

$$\sigma_D = \frac{L_D u'}{D'}\left(1 + \frac{4D'}{u'^2}s\right)^{1/2}.$$ (1.80g)

With the $F(s)$ and the input signal measured, the response signal may be predicted by the techniques mentioned in Section 1.1.6. The parameter values are then determined from a comparison of $C_{calc}^{II}(t)$ and $C_{expt}^{II}(t)$ in the time domain.

Kaguei et al. [20] found, from the input–response signal measurements made in a column ($L = 20.4$ cm and $L_D = 0.4$ to 1.0 cm) at $Re = 0.01$ to 0.17, that the flow rates obtained under the consideration of the dead volume (U_{calc}) agreed well with the experimental data (U_{expt}); the values obtained by ignoring the dead volumes ($U_{calc}^{\dagger}$) were, however, 10–20% lower than U_{expt} or U_{calc}.

The difference between U_{expt} and $U_{calc}^{\dagger}$ may be evaluated by equating the two first moments, obtained separately from Eqs. (1.80) and (1.3), in which the flow rates are distinguished using U_{expt} for Eq. (1.80) (dead volume considered) and $U_{calc}^{\dagger}$ for Eq. (1.3) (dead volume ignored). The result is

$$\frac{U_{expt}}{U_{calc}^{\dagger}} = 1 + \Delta$$ (1.81)

where

$$\Delta = \frac{2L_D}{\epsilon_b L}\left[1 - (1 - \alpha\epsilon_b)\,\nu\right]$$ (1.81a)

$$\nu = \frac{\sinh\beta}{\beta}\exp\left[-\left(2 + \frac{L}{\alpha L_D}\right)\beta\right]$$ (1.81b)

$$\beta = \frac{\epsilon_b U_{expt} L_D}{D'}.$$ (1.81c)

The dead volume effect may be ignored if $\Delta \ll 1$. Otherwise, Eq. (1.80) should be used as the transfer function of the system. Large Δ values result from a column with L_D/L being not very small. As a matter of fact, in such a bed the input signal is closely followed by the response signal.

1.3.2 Adsorption Packed Beds

In a similar way to the preceding section for an inert packed bed, the effect of dead volume in adsorption packed beds can be tested by comparing the adsorption equilibrium constants, K_A (dead volume considered) and $K_A^\dagger$ (dead volume ignored). Again, this is done by equating the two first moments based on Eq. (1.61) (dead volume ignored) and Eq. (1.80) (dead volume considered). Note that in applying Eq. (1.80) in the calculation of the first moment of the adsorption system, σ_B of Eq. (1.80f) should be replaced by $\hat{\sigma}_B$ of Eq. (1.61a). The ratio of $K_A^\dagger$ to K_A is found to be

$$\frac{K_A^\dagger}{K_A} = 1 + \Delta_{ad} \tag{1.82}$$

where

$$\Delta_{ad} = \frac{6L_D}{a\rho_p K_A L R}\left(1 - \left\{1 - \alpha\left[\epsilon_b + \frac{aR}{3}(\epsilon_p + \rho_p K_A)\right]\right\}v\right). \tag{1.82a}$$

Equation (1.82) suggests that if the measurement is made in a short column of adsorbent particles with a low adsorption equilibrium constant, then $K_A^\dagger$ is appreciably different from K_A. In most adsorption columns, however, response signals come far behind the input ones, such that the dead volume may safely be ignored.

Example 1.3

The chromatography data listed in Table 1.5 (see Example 1.2) are those measured using detecting elements (tungsten filaments) installed in the middle of empty sections ($L_D = 0.4$ cm) of the column. Find $\rho_p K_A$ for Run 3, taking into account the dead volume associated with the detecting elements. Assume that $k_a = \infty$ and k_f is estimated from Eq. (4.11), as in Example 1.2.

SOLUTION

At the low flow rate $(Re = 0.30)$, D' is equal to the molecular diffusion coefficient, D_v, for the nitrogen–hydrogen system at $20°C$ and atmospheric pressure. Using the transfer function of Eq. (1.80) (with the modification for adsorption), the response signals are predicted with various values of $\rho_p K_A$, D_{ax} and D_e. Figure 1.25 is the error map of $\rho_p K_A$ versus $\epsilon_b D_{ax}/D_v$ with D_e as a parameter, which corresponds to Figure 1.17 when the dead volume is ignored or when L_D is assumed to be zero.

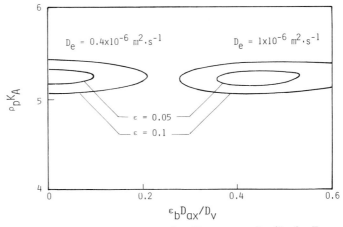

FIGURE 1.25   Error map in the plot of $\rho_p K_A$ versus $\epsilon_b D_{ax}/D_v$ for Example 1.3, when dead volume is considered.

The value of $\rho_p K_A$ with the dead volume considered is then found to be 5.23. In the solution of Example 1.2, the value of $\rho_p K_A^\dagger = 5.29$ is obtained without considering the dead volume. $\rho_p K_A^\dagger$ is about 1% higher than the $\rho_p K_A$. Also, Eq. (1.82a) indicates that $\Delta_{ad} = 0.012$ or 1.2% for Run 3. In any case, the difference between K_A and $K_A^\dagger$ in adsorption chromatography is small and the dead volume may be ignored in the estimation of parameter values.   (*End of Example*)

1.4   Assumption of an Infinite Bed

As mentioned in Section 1.1, a packed bed in which the input and response signals are measured is assumed to be infinite. It is necessary, however, to know the validity of this assumption which has simplified the solution to Eq. (1.1) considerably.

Suppose a response signal is measured in a packed bed at a distance, l, away from the bed exit. It is considered, in general, that if l is short, the shape of the response signal is influenced by the length l, but that the effect diminishes with increasing length. Therefore, a critical bed length exists beyond which the shape of the response signal is independent of bed length. The critical length is affected by flow rate, and, in an adsorption column, also by the magnitude of the adsorption equilibrium constant.

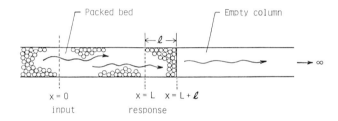

FIGURE 1.26   A column used for examination in an infinite bed.

1.4.1   Packed Beds of Glass Beads

Figure 1.26 shows a packed bed of glass beads connected to an infinitely long empty column. Unsteady-state mass balance equations are as follows:

In the packed bed ($0 < x < L + l$; concentration C)

$$\frac{\partial C}{\partial t} = D_{ax}\frac{\partial^2 C}{\partial x^2} - U\frac{\partial C}{\partial x} \qquad (1.83)$$

In the empty column ($x > L + l$; concentration C')

$$\frac{\partial C'}{\partial t} = D'\frac{\partial^2 C'}{\partial x^2} - u'\frac{\partial C'}{\partial x} \qquad (1.84)$$

with

$$C = C' = 0 \qquad\qquad\qquad\qquad \text{at} \quad t = 0$$

$$C = C' \quad \text{and} \quad \epsilon_b D_{ax} \frac{\partial C}{\partial x} = D' \frac{\partial C'}{\partial x} \qquad \text{at} \quad x = L + l$$

$$C' = 0 \qquad\qquad\qquad\qquad \text{at} \quad x = \infty.$$

The two boundary conditions at the bed exit listed above are the same as those employed in Section 1.3.1.

The transfer function is found to be

$$F(s) = \frac{1 - A \exp\left[-\sigma_B(l/L)\right]}{1 - A \exp\left\{-\sigma_B[1 + (l/L)]\right\}} \exp\left(\lambda_B\right) \tag{1.85}$$

where

$$A = \frac{\lambda_B - \lambda'}{\lambda_B + \sigma_B - \lambda'} \tag{1.85a}$$

$$\lambda' = \frac{Lu'}{2\epsilon_b D_{ax}} \left[1 - \left(1 + \frac{4D's}{u'^2}\right)^{1/2}\right]. \tag{1.85b}$$

λ_B and σ_B are the same as those defined in Section 1.3.1 and $u' = \epsilon_b U$.

The first moment of an impulse response, for example, is then

$$M_1^{II} = \frac{L}{U}(1 + \Lambda_D) \tag{1.86}$$

where

$$\Lambda_D = N_D \left(\frac{D'}{\epsilon_b^2 D_{ax}} - 1\right) \exp\left[-(l/L)/N_D\right][1 - \exp\left(-1/N_D\right)] \tag{1.86a}$$

is a measure of deviation of M_1^{II} from that of an infinite packed bed. The measure of deviation becomes large when l/L decreases and/or N_D increases.

At intermediate and high flow rates, N_D is low, but, with the decrease in flow rate, N_D becomes relatively large and this makes Λ_D large.

The mass dispersion number is rewritten as

$$N_D = \frac{\epsilon_b}{(Sc)(Re)} \frac{D_{ax}}{D_v} \frac{D_p}{L}. \tag{1.87}$$

At low flow rates, the dispersion coefficient, D', in the empty column is identical to the molecular diffusion coefficient, D_v. The axial dispersion coefficient, D_{ax}, in such a packed bed of glass beads is $(0.6 \sim 0.8) \times D_v$ (Eq. 2.29). With the assumptions that $D' = D_v$, $D_{ax} = 0.7 D_v$ and $\epsilon_b = 0.4$, a measure of Λ_D, the deviation from an infinite packed bed, is shown in Figure 1.27 as a function of N_D and l/L.

The packed bed may be assumed to be infinite if the measure of deviation, Λ_D, is less than 0.01. It is then found that a packed bed with $N_D < 0.03$ can be assumed to be infinite if the response measurement is made at $l/L > 0.1$. If $N_D < 0.006$, the criterion for an infinite packed bed is met at $l/L > 0.01$.

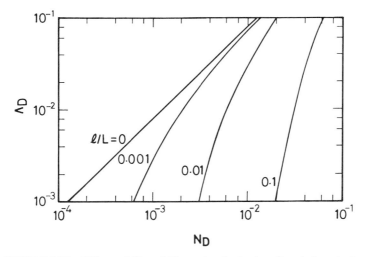

FIGURE 1.27   Effects of l/L and N_D on Λ_D for laminar flow in inert beds.

According to Eq. (1.87), the dispersion number for a laminar gas flow $(Re \lesssim 1)$ with $Sc \simeq 1$ in a bed of $\epsilon_b = 0.4$, for example, is

$$N_D \simeq \frac{0.3\, D_p}{Re\, L}. \tag{1.88}$$

If $L/D_p = 100$, the condition of $N_D < 0.03$ is satisfied at $Re > 0.1$. Similarly, if $L/D_p = 200$, N_D becomes less than 0.03 at $Re > 0.05$.

With the data given in Example 1.1, the effect of l on the response signal is examined. As listed in Table 1.4, we have found from the time domain analysis of the data under the assumption of an infinite bed that $\bar{\tau} = 6.7$ s and $N_D = 0.030$. The molecular diffusion coefficient, D_v, is 0.76×10^{-4} m^2 s^{-1}. With these values, the response curves at various locations are computed. As illustrated in Figure 1.28, if $l/L < 0.01$, the response signals predicted are significantly different from those measured. However, if $l/L > 0.1$, the signals computed do not differ appreciably from the signal measured. Therefore, we find that, as far as the packed bed of Example 1.1 is concerned, the bed of glass beads may be assumed to be infinite, if the response signal is measured at $l/L > 0.1$.

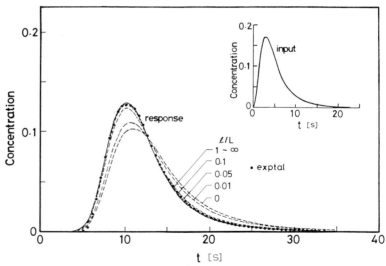

FIGURE 1.28   Effect of l/L on response curves predicted.

1.4.2   Adsorption Packed Beds

A similar computation is made for an adsorption bed. The bed end is again assumed to be connected to an infinitely long empty column. The system is then described by Eqs. (1.57)-(1.60) and (1.84) with the same conditions as listed in the preceding section.

The transfer function is

$$F(s) = \frac{1 - \hat{A} \exp\left[-\hat{\sigma}_B(l/L)\right]}{1 - \hat{A} \exp\left\{-\hat{\sigma}_B[1 + (l/L)]\right\}} \exp(\hat{\lambda}_B) \qquad (1.89)$$

where

$$\hat{A} = \frac{\hat{\lambda}_B - \lambda'}{\hat{\lambda}_B + \hat{\sigma}_B - \lambda'} \qquad (1.89a)$$

$$\hat{\lambda}_B = \frac{1}{2}\left(\frac{LU}{D_{ax}} - \hat{\sigma}_B\right) \qquad (1.89b)$$

and λ' is defined by Eq. (1.85b) and $\hat{\sigma}_B$ by Eq. (1.61a).

The first moment of the impulse response is then

$$M_1^{II} = \frac{L}{U}(1 + \delta_0)(1 + \Lambda_{ad}) \qquad (1.90)$$

where

$$\Lambda_{ad} = N_D \Gamma \exp\left[-(l/L)/N_D\right]\left[1 - \exp(-1/N_D)\right] \qquad (1.90a)$$

and

$$\Gamma = \frac{D'}{\epsilon_b^2 D_{ax}(1 + \delta_0)} - 1 \qquad (1.90b)$$

and δ_0 is defined by Eq. (1.63a).

If $(\epsilon_p + \rho_p K_A) = 0$, Λ_{ad} becomes Λ_D for inert bed. This Λ_{ad} is a measure of deviation from the infinite adsorption column. The Λ_{ad} is relatively large at low flow rates. Assuming that $D' = D_v$, $D_{ax} = 0.7 D_v$ and

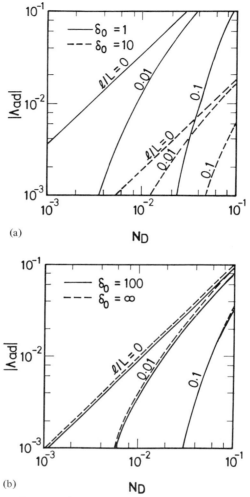

FIGURE 1.29   Effects of l/L, N_D and δ_0 on $|\Lambda_{ad}|$ for laminar flow in adsorption beds: (a) $\delta_0 = 1$ and 10; (b) $\delta_0 = 100$ and ∞.

$\epsilon_b = 0.4$, similar to the preceding section, Λ_{ad} for an adsorption column at low flow rates is evaluated as a function of N_D, δ_0 and l/L. Figures 1.29(a) and (b) show the relationships between $|\Lambda_{ad}|$, δ_0 and N_D at $l/L = 0$, 0.01 and 0.1. Again, the criterion for an infinite bed may be given by $|\Lambda_{ad}| = 0.01$.

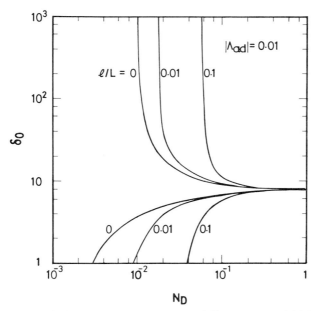

FIGURE 1.30   Relationships between N_D, δ_0 and l/L at $|\Lambda_{ad}| = 0.01$ for laminar flow in adsorption beds.

Figure 1.30 shows the relationship between N_D, δ_0 and l/L, which satisfies $|\Lambda_{ad}| = 0.01$. When N_D and δ_0 are known, we can easily see where in the bed the response signals should be measured in order to satisfy the assumption of an infinite bed length.

The impulse responses at $Re = 0.05$, for example, are computed with the following data (the same as Example 1.2):

$$\epsilon_b = 0.38$$

$$\epsilon_p = 0.59$$

$$D_p = 0.2 \text{ cm}$$

$$D_e = 0.63 \times 10^{-6} \text{ m}^2 \text{ s}^{-1}$$

$$D_v = 0.76 \times 10^{-4} \text{ m}^2 \text{ s}^{-1}$$

$$\epsilon_b D_{ax}/D_v = 0.24$$

$$L = 20.4 \text{ cm}$$

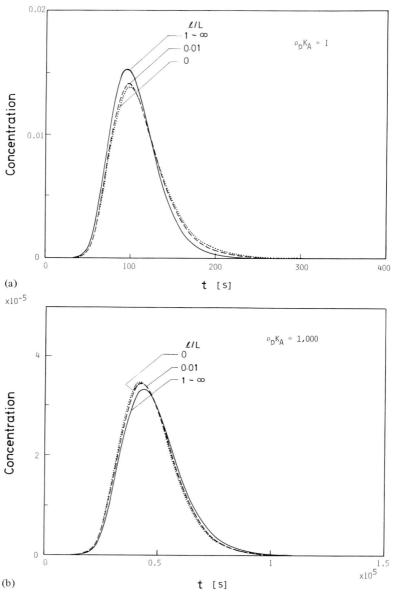

FIGURE 1.31   Effect of $\rho_p K_A$ on impulse responses for Example 1.2 with $Re = 0.05$: (a) $\rho_p K_A = 1$; (b) $\rho_p K_A = 1000$.

k_f from Eq. (4.11)

$$k_a = \infty.$$

As shown in Figure 1.31, the impulse response at $l/L = 1$ to ∞ differs considerably from those at $l/L = 0$ and 0.01, when $\rho_p K_A$ is small. It is interesting to note that the peak of the impulse response with $l/L = 1$ to ∞ is higher than those with $l/L = 0$ and 0.01 as long as $\rho_p K_A$ is small. When $\rho_p K_A$ is high, however, the peak of the impulse response with $l/L = 0$ becomes the highest. This comes from the fact that with an increase in δ_0 or $\rho_p K_A$, the value of Γ turns from positive to negative. The first moment of the impulse response then decreases and the response peak gets high.

The moment equations, such as Eqs. (1.63) and (1.64), may be derived from the transfer function in a digital computer. The computer programs for the derivations are listed in Appendix C.

REFERENCES

[1] K. Østergaard and M. L. Michelsen, *Can. J. Chem. Eng.* **47**, 107 (1969).
[2] S. K. Gangwal, R. R. Hudgins, A. W. Bryson and P. L. Silveston, *Can. J. Chem. Eng.* **49**, 113 (1971).
[3] W. C. Clements, *Chem. Eng. Sci.* **24**, 957 (1969).
[4] A. S. Anderssen and E. T. White, *Chem. Eng. Sci.* **25**, 1015 (1970).
[5] A. S. Anderssen and E. T. White, *Chem. Eng. Sci.* **26**, 1203 (1971).
[6] H. J. Wolff, K. H. Radeke and D. Gelbin, *Chem. Eng. Sci.* **34**, 101 (1979).
[7] M. J. Hopkins, A. J. Sheppard and P. Eisenklam, *Chem. Eng. Sci.* **24**, 1131 (1969).
[8] M. Kubin, *Collec. Czech. Chem. Commun.* **30**, 1104; 2900 (1965).
[9] E. Kucera, *J. Chromatography* **19**, 237 (1965).
[10] P. Schneider and J. M. Smith, *AIChE J.* **14**, 762 (1968).
[11] N. Wakao, K. Tanaka and H. Nagai, *Chem. Eng. Sci.* **31**, 1109 (1976).
[12] N. Wakao, S. Kaguei and J. M. Smith, *J. Chem. Eng. Japan* **12**, 481 (1979).
[13] C. R. Antonson and J. S. Dranoff, *Chem. Eng. Prog. Symp. Ser.* **65** (No. 96), 27 (1969).
[14] E. Ruckenstein, A. S. Vaidyanathan and G. R. Youngquist, *Chem. Eng. Sci.* **26**, 1305 (1971).
[15] N. Hashimoto and J. M. Smith, *Ind. Eng. Chem. Fund.* **12**, 353 (1973).
[16] K. Kawazoe and Y. Takeuchi, *J. Chem. Eng. Japan* **7**, 431 (1974).
[17] P. Schneider and J. M. Smith, *AIChE J.* **14**, 886 (1968).
[18] K. Kawazoe, M. Suzuki and K. Chihara, *J. Chem. Eng. Japan* **7**, 151 (1974).
[19] L. K. Lee and D. M. Ruthven, *Can. J. Chem. Eng.* **57**, 65 (1979).
[20] S. Kaguei, K. Matsumoto and N. Wakao, *Chem. Eng. Sci.* **35**, 1809 (1980).

2 Fluid Dispersion Coefficients

It is well recognized that conversion in a chemical reactor depends largely on the degree of fluid dispersion in the reactor. Axial fluid dispersion coefficients have been obtained mainly from tracer injection measurements, as discussed in Section 1.1. In packed beds of non-porous particles such as glass beads, no tracer species penetrates into the particles, and it is considered that axial dispersion of the tracer, while flowing in the bed, is described by Eq. (1.1).

However, in the case of mass transfer taking place inside the particle, the intraparticle concentration is, as mentioned in Section 1.2 for an adsorption bed, usually assumed to have radial symmetry. This convention has been introduced to simplify the solution to the problem. Dispersion itself is a hydrodynamic phenomenon which occurs while the fluid is flowing in the interstitial space of a packed bed. Fluid dispersion should, therefore, be independent of what is occurring inside the particle. However, the question is whether the assumption of concentric intraparticle concentration can describe the mass transfer phenomenon sufficiently enough or not. If not, this may superficially alter the value of the dispersion coefficient which appears in the fundamental equations derived under the assumption of radially symmetric concentration.

In this chapter, the theoretical treatment of the effect of dispersion on chemical conversion, the significance of fluid dispersion coefficients and their evaluation in reactive, non-reactive and adsorption packed bed systems are discussed.

2.1 Effect of Dispersion on Conversion

In a continuous flow reactor in which a chemical reaction is taking place under steady-state conditions, the mass balance equation for a reacting

72

species is:

$$U \frac{dC}{dx} - D_{ax} \frac{d^2C}{dx^2} - \frac{r_x}{\epsilon_b} = 0 \tag{2.1}$$

where r_x is the reaction rate per unit volume of reactor and defined as the production rate of the species under consideration. For a reactant disappearing in the reactor, the reaction rate is negative.

Equation (2.1) is analytically soluble only when the reaction rate is zeroth-order or first-order with respect to the concentration of the reactant. The reaction rate, for example, with first-order kinetics is:

$$r_x = -KC \tag{2.2}$$

where K is the reaction rate constant. When the reaction proceeds only in a fluid phase, K is the intrinsic rate constant of the homogeneous chemical reaction. If, however, the reaction takes place in a porous catalyst particle, K is then an overall rate constant, which depends, in general, not only upon the intrinsic chemical reaction rate, but also on the mass transfer rates both at the particle surface and inside the particle.

If the reaction is not zeroth-order or first-order, Eq. (2.1) should be solved numerically. In any case, solving Eq. (2.1) requires two axial boundary conditions. The Danckwerts boundary conditions [1] are widely used where:

$$UC_{in} = UC - D_{ax} \frac{dC}{dx} \qquad \text{at } x = 0 \text{ (inlet)} \tag{2.3}$$

$$\frac{dC}{dx} = 0 \qquad \text{at } x = L \text{ (exit)} \tag{2.4}$$

where C_{in} is the concentration of the reactant in the fluid flowing into the reactor. The inlet condition, Eq. (2.3), has been derived under the assumption that no fluid dispersion occurs before the reactor, in which the fluid is flowing in the dispersed plug flow mode. The exit condition is established intuitively.

With the reaction rate given by Eq. (2.2) and the Danckwerts boundary conditions, the solution to Eq. (2.1) under isothermal conditions is:

$$\frac{C}{C_{in}} = A \exp\left(\frac{x}{2LN_D}\right)\left\{(1+B)\exp\left[\frac{B\left(1-\frac{x}{L}\right)}{2N_D}\right]\right.$$

$$\left. - (1-B)\exp\left[-\frac{B\left(1-\frac{x}{L}\right)}{2N_D}\right]\right\} \tag{2.5}$$

where

$$A = \frac{2}{(1+B)^2 \exp\left(\frac{B}{2N_D}\right) - (1-B)^2 \exp\left(-\frac{B}{2N_D}\right)} \tag{2.5a}$$

$$B = \left(1 + 4N_D \frac{K\bar{\tau}}{\epsilon_b}\right)^{1/2} \tag{2.5b}$$

$$N_D = \frac{D_{ax}}{LU} \tag{2.5c}$$

$$\bar{\tau} = \frac{L}{U}. \tag{2.5d}$$

The Danckwerts boundary conditions were first verified by Wehner and Wilhelm [2], under the condition that C_{in} was the concentration at $x \ll 0$. In their verification, they considered a column consisting of three sections, as shown in Figure 2.1. Sections 1 and 3 are inert zones in which no reaction occurs and are assumed to be semi-infinite in length. Between them is a reaction zone (Section 2), where a first-order reaction is proceeding.

Under steady-state conditions, mass balance equations for the reactant in the three separate sections are:

In Section 1 ($x < 0$; concentration C')

$$u \frac{dC'}{dx} - E'_{ax} \frac{d^2C'}{dx^2} = 0 \tag{2.6}$$

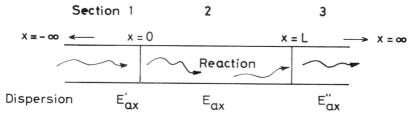

FIGURE 2.1   A column composed of three sections.

In Section 2 (0<x<L; concentration C)

$$u\frac{dC}{dx} - E_{ax}\frac{d^2C}{dx^2} + KC = 0 \qquad (2.7)$$

In Section 3 (x >L; concentration C")

$$u\frac{dC''}{dx} - E''_{ax}\frac{d^2C''}{dx^2} = 0. \qquad (2.8)$$

The fluid velocity, u, and the axial fluid dispersion coefficients, E'_{ax}, E_{ax} and E''_{ax}, are all based on the cross-section of the column.

The boundary conditions are:

(a) $C' = C_{in}$ at $x = -\infty$

(b) $uC' - E'_{ax}\dfrac{dC'}{dx} = uC - E_{ax}\dfrac{dC}{dx}$ at $x = 0$

(c) $C' = C$

(d) $uC - E_{ax}\dfrac{dC}{dx} = uC'' - E''_{ax}\dfrac{dC''}{dx}$ at $x = L$

(e) $C = C''$

(f) $C'' = $ finite   at $x = \infty$.

Conditions (b) and (d) result from the conservation of reactant at the bed inlet and outlet. Conditions (c) and (e) come from the intuitive argument that the concentrations should be the same at the intersections.

The solutions are:

$$\frac{C_{in} - C'}{C_{in} - C(0)} = \exp\left(\frac{x}{LN_D'}\right) \qquad \text{for } x \leqslant 0 \qquad (2.9)$$

where

$$C(0) = C_{in} A \left[(1+B) \exp\left(\frac{B}{2N_D}\right) - (1-B) \exp\left(-\frac{B}{2N_D}\right)\right] \quad (2.9a)$$

$$N_D' = \frac{E_{ax}'}{Lu} \qquad (2.9b)$$

and

$$\frac{C}{C_{in}} = \text{Eq. (2.5)} \qquad \text{for } 0 \leqslant x \leqslant L$$

$$\frac{C''}{C_{in}} = \frac{C_{exit}}{C_{in}} = 2AB \exp\left(\frac{1}{2N_D}\right) \qquad \text{for } x \geqslant L. \qquad (2.10)$$

The axial fluid dispersion coefficient, D_{ax}, based on unit void area, is equal to E_{ax}/ϵ_b, and the interstitial fluid velocity, U, is equal to u/ϵ_b. Therefore, the mass dispersion number for Section 2, for example, is

$$N_D = \frac{E_{ax}}{Lu} = \frac{D_{ax}}{LU}. \qquad (2.11)$$

Wehner and Wilhelm [2] found that the concentration profile in Section 2 was identical to that predicted from the solution obtained under the Danckwerts boundary conditions. It is also found that N_D' has no effect on the concentration profiles in Sections 2 and 3. As depicted in Figure 2.2, if N_D' is zero, the concentration profile at $x = 0-$ approaches that of a step function. With increasing N_D', the concentration gradient decreases. In any case, the concentration profile in Section 2 depends only upon N_D. Also, the concentration gradient is zero at $x = L$. The fluid dispersion in Section 3 has no effect on concentration profile in any section.

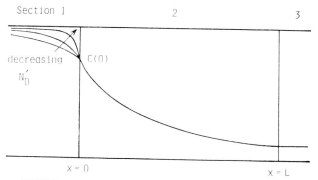

FIGURE 2.2   Concentration profiles in Sections 1, 2 and 3.

Equation (2.10) shows that the exit concentrations, C_{exit}, in the following two extremes are:

For a plug-flow reactor (with $N_D = 0$)

$$\frac{C_{exit}}{C_{in}} = \exp\left(-\frac{K\bar{\tau}}{\epsilon_b}\right)$$

$$= \exp\left(-\frac{KV}{F_v}\right) \qquad (2.12)$$

For a continuous stirred tank reactor (with $N_D = \infty$)

$$\frac{C_{exit}}{C_{in}} = \frac{1}{1 + \dfrac{K\bar{\tau}}{\epsilon_b}}$$

$$= \frac{1}{1 + \dfrac{KV}{F_v}} \qquad (2.13)$$

where $\bar{\tau}$ is the mean residence time, V is the reactor volume and F_v is the volumetric flow rate; $\bar{\tau} = \epsilon_b V/F_v$.

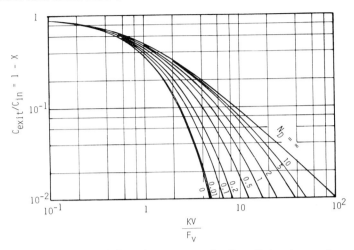

Figure 2.3 is a graphical presentation of Eq. (2.10). The conversion, X, is defined by

$$X = 1 - \frac{C_{exit}}{C_{in}}. \tag{2.14}$$

Obviously, a plug flow reactor requires the least reactor volume to attain a given conversion. With an increase in N_D, the reactor volume becomes larger, and the maximum is that of a single continuous stirred tank reactor.

Since the first confirmation by Wehner and Wilhelm [2], the Danckwerts boundary conditions have been successively re-examined and employed for different reaction schemes, both isothermal and non-isothermal, by van der Laan [3], Fan and Bailie [4], Bischoff [5], Bischoff and Levenspiel [6], Fan and Ahn [7], Carberry and Wendel [8], Liu and Amundson [9], Hofmann and Astheimer [10], Schmeal and Amundson [11], Mears [12], Wen and Fan [13] and Chang et al. [14]. The criteria, in applying the Danckwerts boundary conditions, are given by Gunn [15] as: the Danckwerts conditions are realized only when the tracer imposed moves to some extent against the main direction of flow in the reactor, i.e. when diffusional dispersion is controlling. For dynamic study,

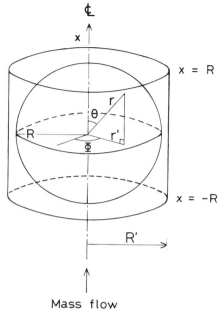

Mass flow

FIGURE 2.4   Single cell and coordinates.

however, appropriate boundary conditions other than those of Danckwerts have often been employed. More detailed information is given by Carbonell [16], for example.

2.2   Fluid Dispersion Coefficients in a Reacting System

The fluid dispersion coefficient needed for the design of a packed bed reactor is that for the reacting species in the fluid. It had long been assumed that the dispersion coefficient for a reacting species was the same as that in an inert system. However, Wakao *et al.* [17] have studied the dispersion coefficient for a reacting species, and have found it to be considerably different from that under inert conditions. First, their study of the dispersion coefficient at zero flow rate is outlined below.

Suppose a short cylinder as shown in Figure 2.4, consisting of a spherical catalyst particle and a stagnant fluid, is the smallest element or unit cell of a multiparticle system. The height of the cylinder is assumed to equal the sphere diameter $2R$, and the radius R' of the cylinder is $1.05R$, the void volume fraction in the cell is then 0.4. Also, no mass transfer is assumed to occur across the side of the cylinder.

When an isothermal, first-order, irreversible, catalytic reaction proceeds in the sphere under steady-state conditions, mass balance equations for the reactant give

$$D_v \nabla^2 C^* = 0 \qquad \text{for } r > R \qquad (2.15)$$

$$D_e \nabla^2 c^* - k_x c^* = 0 \qquad \text{for } 0 < r < R \qquad (2.16)$$

where

$$\nabla^2 = \frac{1}{r^2} \frac{\partial}{\partial r}\left(r^2 \frac{\partial}{\partial r}\right) + \frac{1}{r^2}\left[\frac{1}{\sin\theta} \frac{\partial}{\partial\theta}\left(\sin\theta \frac{\partial}{\partial\theta}\right) + \frac{1}{\sin^2\theta} \frac{\partial^2}{\partial\Phi^2}\right] \quad (2.16a)$$

with

$$C^* = c^* \quad \text{and} \quad D_v \frac{\partial C^*}{\partial r} = D_e \frac{\partial c^*}{\partial r} \qquad \text{at } r = R$$

$$\frac{\partial C^*}{\partial r'} = 0 \qquad \text{at } r' = R'$$

where

$C^* =$ concentration in stagnant fluid
$c^* =$ concentration in a particle
$D_e =$ intraparticle effective diffusivity
$D_v =$ molecular diffusion coefficient
$k_x =$ first-order reaction rate constant
$r =$ radial distance in spherical coordinates
$r' =$ radial distance in cylindrical coordinates.

The complete solution to Eqs. (2.15) and (2.16) is analytically difficult. Therefore, instead, let us assume that what happens in the unit cell (Cell A) may be considered to take place in two separate unit cells (Cell B and Cell C) as shown in Figure 2.5. Suppose the following boundary conditions are imposed:

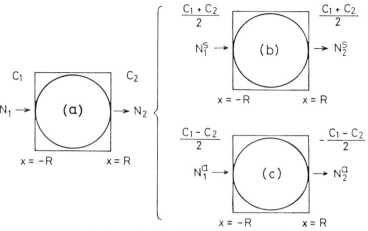

FIGURE 2.5   Equivalence of cell models: (a) Case A; (b) Case B; (c) Case C.

Cell A (Figure 2.5a, C^* and c^* are the concentrations in the stagnant fluid and particle, respectively).

$$\left.\begin{aligned} C^* &= C_1 \qquad \text{at } x = -R \\ C^* &= C_2 \qquad \text{at } x = R \end{aligned}\right\} \qquad (2.17)$$

Cell B (Figure 2.5b, C^{*s} and c^{*s} are the concentrations in the stagnant fluid and particle, respectively).

$$\left.\begin{aligned} C^{*s} &= \frac{C_1 + C_2}{2} \qquad \text{at } x = -R \\ C^{*s} &= \frac{C_1 + C_2}{2} \qquad \text{at } x = R \end{aligned}\right\} \qquad (2.18)$$

Cell C (Figure 2.5c, C^{*a} and c^{*a} are the concentrations in the stagnant fluid and particle, respectively).

$$\left.\begin{aligned} C^{*a} &= \frac{C_1 - C_2}{2} \qquad \text{at } x = -R \\ C^{*a} &= -\frac{C_1 - C_2}{2} \qquad \text{at } x = R. \end{aligned}\right\} \qquad (2.19)$$

Similar to Cell A, it is assumed in Cells B and C that no mass transfer occurs across the sides of the cylinders. The solutions in Cell A are then related to those of Cells B and C by the following equations:

$$C^* = C^{*s} + C^{*a} \tag{2.20}$$

$$c^* = c^{*s} + c^{*a}. \tag{2.21}$$

In Cell B: the boundary conditions given by Eq. (2.18) indicate that the concentrations, C^{*s} and c^{*s}, are even functions of x or symmetric with respect to x; hence the diffusion rates at both ends of the cell are equal but in opposite directions, i.e. $N_1^s = -N_2^s$. In other words, there is no net diffusion flux of the species passing across the cell.

In Cell C: to the contrary, the boundary conditions, Eq. (2.19), show that the concentrations, C^{*a} and c^{*a}, are odd functions of x or antisymmetric with respect to x; the diffusion rates are, therefore, equal and in the same direction, i.e. $N_1^a = N_2^a$. Thus, the number of moles of reacting species depleted due to chemical reaction in the left hemisphere of the solid particle is compensated by the appearance of an equal amount produced in the right hemisphere. As a result, there is no net change in the total number of moles of reacting species in the particle.

The boundary concentrations and the diffusion rates of the proposed cell models are summarized in Table 2.1.

The above analysis shows that there is no net change in the number of moles of reacting species in the solid particle in Cell C, whereas, in Cell B, the net diffusion flux is zero. Therefore, the changes in the number of

TABLE 2.1
Characteristics of cell models.

Cell	Concentrations Fluid	Concentrations Solid	Boundary concentrations $x = -R$	Boundary concentrations $x = R$	Diffusion rates across end-faces $x = -R$	Diffusion rates across end-faces $x = R$
A	C^*	c^*	C_1	C_2	N_1	N_2
B	C^{*s}	c^{*s}	$\dfrac{C_1 + C_2}{2}$	$\dfrac{C_1 + C_2}{2}$	$N_1^s = \dfrac{N_1 - N_2}{2}$	$N_2^s = -\dfrac{N_1 - N_2}{2}$
C	C^{*a}	c^{*a}	$\dfrac{C_1 - C_2}{2}$	$-\dfrac{C_1 - C_2}{2}$	$N_1^a = \dfrac{N_1 + N_2}{2}$	$N_2^a = \dfrac{N_1 + N_2}{2}$

moles of reacting species in the particle and the diffusion rate in Cell A must correspond to those in Cell B and Cell C, respectively. Hence, the diffusion rate across Cell A can be determined from that of Cell C and vice versa. The diffusion coefficient of the reacting species in Cell A can, therefore, be assumed to be the same as that in Cell C. The effective diffusion coefficient, E^o, of the reacting species in Cell C is defined as:

$$\frac{N_1 + N_2}{2} = (\pi R'^2) E^o \left(\frac{C_1 - C_2}{2R}\right). \tag{2.22}$$

Using a grid network for Cell A, steady-state concentrations at nodal points are computed for $D_e/D_v = 10^{-3}$ to 10, and the Jüttner modulus (see Section 3.2) $\phi = R(k_x/D_e)^{1/2} = 0$ to ∞. The predicted concentration profiles are illustrated for the cases where $\phi = 0.5$ and 5 in Figures 2.6(a)–(c). The catalyst effectiveness factor (see Chapter 3) is close to unity at $\phi = 0.5$, and is 0.48 at $\phi = 5$. In either case, Figure 2.6 reveals that the intraparticle concentration is not radially symmetric, especially, at lower ϕ values.

The diffusion rates, N_1 and N_2, in Cell A are evaluated with the calculated concentration gradients and the grid conductances at both ends of

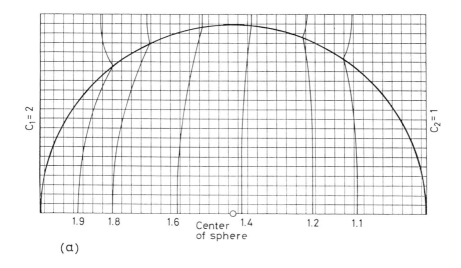

(a)

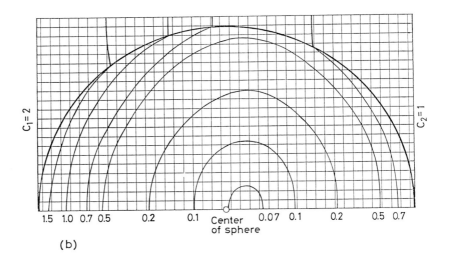

(b)

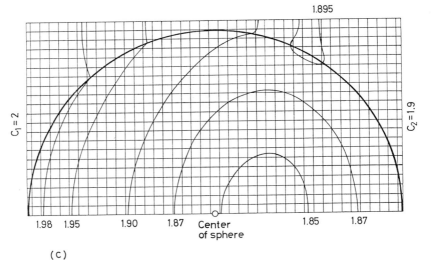

(c)

FIGURE 2.6   Calculated concentration profiles of reacting species in the cell with $D_e = 0.08 \times 10^{-4}$ m^2 s^{-1} and $D_v = 0.7 \times 10^{-4}$ m^2 s^{-1}: (a) $C_1 = 2$, $C_2 = 1$ and $\phi = 0.5$; (b) $C_1 = 2$, $C_2 = 1$ and $\phi = 5$; (c) $C_1 = 2$, $C_2 = 1.9$ and $\phi = 0.5$.

the cylinder. With these diffusion rates, and given C_1 and C_2, the effective diffusion coefficients are evaluated from Eq. (2.22). The coefficient values obtained are then plotted in Figure 2.7 as E^o/D_v versus D_e/D_v with ϕ as a parameter.

As depicted in Figure 2.7, the E^o values are surprisingly large, and also, they depend on the Jüttner modulus, ϕ, or catalyst effectiveness factor, as well as on the fluid and intraparticle diffusivities. It should be noted that concentration profiles depend upon the assumed values of C_1 and C_2 (refer to Figures 2.6a and c), but the values of E^o defined by Eq. (2.22) are independent of the boundary concentration values as long as the reaction is first-order.

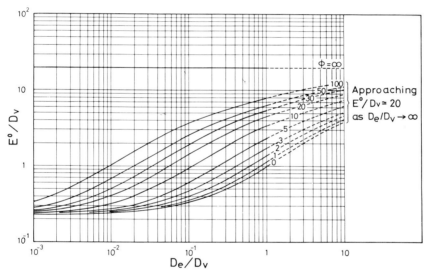

FIGURE 2.7   Effective diffusion coefficients in a quiescent bed under first-order reaction conditions.

In Figure 2.7, the curve for $\phi = 0$ corresponds to the diffusional dispersion when no chemical reaction takes place in a bed of porous particles. The curve for $\phi = 0$ approaches $(E^o)_{inert}/D_v = 0.23$ when D_e/D_v goes to zero (non-porous particles). When $\phi = \infty$, chemical reaction is fast and proceeds only at the surface, so that the reactants are at zero concentrations (for irreversible reaction) or at equilibrium concentrations (for a reversible reaction) on the catalyst particle surface as well as within the particle. Under such conditions, the contribution due to diffusional dispersion is large as shown in Figure 2.7: $E^{\circ}/D_v = 20$.

The fluid dispersion coefficient is considered to consist of diffusional and turbulent contributions. The turbulent contribution is well recognized and, in terms of the Peclet number, is given as:

$$(Pe_r)_{mixing} = D_p U/(D_r)_{mixing} \simeq 10 \qquad (2.23)$$

$$(Pe_{ax})_{mixing} = D_p U/(D_{ax})_{mixing} \simeq 2. \qquad (2.24)$$

The diffusional contribution corresponds to the effective diffusion coefficient at zero flow rate and is isotropic. In the range of laminar flow, $Re < 1$, the dispersion coefficient only consists of a diffusional contribution. At Reynolds number greater than 5, flow is turbulent. Hence, the radial dispersion coefficients are given by the following equations:

$$
\left.
\begin{aligned}
D_r &= \frac{E^{\circ}}{\epsilon_b} & \text{for } Re < 1 \\[2mm]
&= \frac{E^{\circ}}{\epsilon_b} + 0.1 D_p U & \text{for } Re > 5.
\end{aligned}
\right\} \qquad (2.25)
$$

Similarly, the axial dispersion coefficients are

$$
\left.
\begin{aligned}
D_{ax} &= \frac{E^{\circ}}{\epsilon_b} & \text{for } Re < 1 \\[2mm]
&= \frac{E^{\circ}}{\epsilon_b} + 0.5 D_p U & \text{for } Re > 5.
\end{aligned}
\right\} \qquad (2.26)
$$

The diffusional contribution, obtained by Edwards and Richardson [18], Evans and Kenney [19], Suzuki and Smith [20] and Wakao *et al.* [21] from dispersion measurements of non-adsorbing species in packed beds of non-porous particles, has been expressed as:

$$\frac{(E^{\circ})_{inert}}{D_v} = (0.6 \sim 0.8) \, \epsilon_b. \qquad (2.27)$$

Therefore, if a particle (both external and internal surfaces) is involved in neither a reaction nor in a mass transfer process, the radial and axial dispersion coefficients in such an inert bed are

$$(D_r)_{inert} = (0.6 \sim 0.8) D_v \qquad \text{for } Re < 1$$
$$= (0.6 \sim 0.8) D_v + 0.1 D_p U \qquad \text{for } Re > 5 \qquad (2.28)$$

$$(D_{ax})_{inert} = (0.6 \sim 0.8) D_v \qquad \text{for } Re < 1$$
$$= (0.6 \sim 0.8) D_v + 0.5 D_p U \qquad \text{for } Re > 5. \qquad (2.29)$$

In the past the radial and axial dispersion coefficients for reacting species have been assumed simply to be identical to the corresponding dispersion coefficients in an inert bed. However, as long as the turbulent contributions are not dominant in Eqs. (2.25) and (2.26) it has been shown that the dispersion coefficients of reacting species are significantly different from those under inert conditions. It should be noted that if a chemical reaction occurs homogeneously only in the fluid phase in a packed bed, the dispersion coefficients for the reacting species are given by Eqs. (2.28) and (2.29).

2.3 Fluid Dispersion Coefficients in Adsorption Beds

Let us examine again the dispersion coefficient for an adsorbing gaseous species at zero flow rate based on the cell models (Figure 2.4). The concentration of the adsorbing species in the gas phase, C^*, is governed by the following equation:

$$\frac{\partial C^*}{\partial t} = D_v \nabla^2 C^* \qquad \text{for } r > R \qquad (2.30)$$

where ∇^2 is given by Eq. (2.16a). The intraparticle concentration, c^*, is expressed by Eq. (1.73).

In physical adsorption, the adsorption rate constant is so large that it may be assumed that $k_a = \infty$ (refer to Section 1.2.2). Equation (1.74) then gives $c_{ad}^* = K_A c^*$. Substituting this into Eq. (1.73), we have

$$(\epsilon_p + \rho_p K_A) \frac{\partial c^*}{\partial t} = D_e \nabla^2 c^* \qquad \text{for } r < R. \qquad (2.31)$$

Equations (2.30) and (2.31) are numerically solved under the following conditions:

$$C^* = c^* = 0 \qquad \text{at } t = 0$$

$$C^* = 1 \qquad \text{at } x = -R$$

$$C^* = 0 \qquad \text{at } x = R$$

$$C^* = c^* \quad \text{and} \quad D_v \frac{\partial C^*}{\partial r} = D_e \frac{\partial c^*}{\partial r} \qquad \text{at } r = R.$$

C^* and c^* are functions of three variables, x, $r' = r \sin \theta$, and t. The diffusion rate, n_x, passing axially through a cross-sectional area of the unit cell is defined as:

$$n_x = -2\pi \int_0^{R'} D \frac{\partial \Theta}{\partial x} r' \, dr' \qquad (2.32)$$

where

$$D = D_e, \Theta = c^* \qquad \text{for } 0 < r < R$$

$$D = D_v, \Theta = C^* \qquad \text{for } r > R.$$

The average diffusion rate, N, in the unit cell is expressed in terms of the transient effective diffusion coefficient, $E^o(t)$, as:

$$N = \frac{1}{2R} \int_{-R}^{R} n_x \, dx$$

$$= (\pi R'^2) E^o(t) \frac{\Delta \Theta}{2R} \qquad (2.33)$$

where $\Delta \Theta = C^*_{x=-R} - C^*_{x=R}$. From Eqs. (2.32) and (2.33), we have

$$E^{\circ}(t) = \frac{2R}{\pi R'^2 \Delta\Theta} N$$

$$= -\frac{2}{R'^2 \Delta\Theta} \int\limits_{-R}^{R} dx \int\limits_{0}^{R'} D \frac{\partial\Theta}{\partial x} r' dr'. \tag{2.34}$$

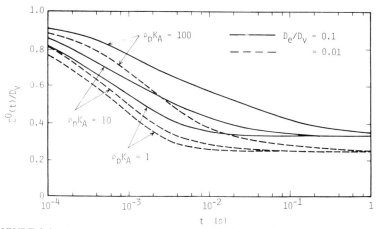

FIGURE 2.8   Transient effective diffusion coefficients under adsorption conditions.

As an illustration, the calculations made with the following data are shown in Figure 2.8:

$$R = 0.1 \text{ cm}$$
$$R' = 1.05R$$
$$D_v = 0.8 \times 10^{-4} \text{ m}^2 \text{ s}^{-1}$$
$$\epsilon_p = 0.5$$
$$\rho_p K_A = 1, 10 \text{ and } 100$$
$$D_e/D_v = 0.01 \text{ and } 0.1.$$

As shown, the effective diffusion coefficients, $E^{\circ}(t)$, are high at $t = 0$, but decrease rapidly with increasing time until steady-state values are reached. The decrease becomes more gradual with an increase in $\rho_p K_A$, but the time to reach the steady-state values is always very short. Moreover, the steady-state values of E^0/D_v depend only upon D_e/D_v. The steady-state values of E^0/D_v, attained by the adsorbing species, are found to be the same as those under inert conditions.

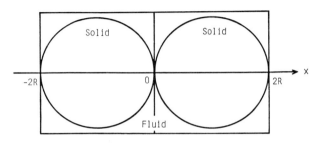

FIGURE 2.9   Two unit cells in contact.

For two cells in contact [writing $C^*(II)$ and $c^*(II)$] as shown in Figure 2.9; if the initial and boundary conditions are chosen as:

$$C^*(II) = c^*(II) = 0 \qquad \text{at } t = 0$$

$$C^*(II) = 1 \qquad \text{at } x = -2R$$

$$C^*(II) = 0 \qquad \text{at } x = 2R.$$

The solutions $C^*(II)$ and $c^*(II)$ are expressed, similar to Eqs. (2.20) and (2.21), as follows:

$$C^*(II) = C^*(II)^s + C^*(II)^a \qquad (2.35)$$

and

$$c^*(II) = c^*(II)^s + c^*(II)^a \qquad (2.36)$$

where $C^*(II)^s$ and $c^*(II)^s$ are symmetric with respect to x and are the solutions obtained under the conditions:

$$C^*(II)^s = c^*(II)^s = 0 \qquad \text{at } t = 0$$

$$C^*(II)^s = 1/2 \qquad \text{at } x = -2R$$

$$C^*(II)^s = 1/2 \qquad \text{at } x = 2R$$

and $C^*(II)^a$ and $c^*(II)^a$ are antisymmetric with respect to x and are the solutions under the conditions:

$$C^*(II)^a = c^*(II)^a = 0 \qquad \text{at } t = 0$$

$$C^*(II)^a = 1/2 \qquad \text{at } x = -2R$$

$$C^*(II)^a = -1/2 \qquad \text{at } x = 2R$$

or

$$C^*(II)^a = 0 \qquad \text{at } x = 0.$$

Similar to Eq. (2.34), the effective diffusion coefficient, $E^o(II)$, for the two cells in contact is

$$E^o(II) = -\frac{2}{R'^2 \Delta\Theta(II)} \int_{-2R}^{2R} dx \int_0^{R'} D \frac{\partial[\Theta(II)^s + \Theta(II)^a]}{\partial x} r' \, dr' \qquad (2.37)$$

where

$$\Delta\Theta(II) = C^*(II)_{x=-2R} - C^*(II)_{x=2R} = 1$$

$$\Theta(II)^s = c^*(II)^s, \Theta(II)^a = c^*(II)^a, D = D_e \qquad \text{for } 0 < r < R$$

$$\Theta(II)^s = C^*(II)^s, \Theta(II)^a = C^*(II)^a, D = D_v \qquad \text{for } r > R.$$

Since $\partial\Theta(II)^s/\partial x$ is antisymmetric and $\partial\Theta(II)^a/\partial x$ is symmetric with respect to x, Eq. (2.37) becomes

$$E^o(II) = -\frac{4}{R'^2} \int_{-2R}^0 dx \int_0^{R'} D \frac{\partial\Theta(II)^a}{\partial x} r' \, dr'. \qquad (2.38)$$

As mentioned already, the concentrations at $x = -2R$ and $x = 0$ are $\Theta(II)^a = 1/2$ and 0, respectively, and consequently $\Delta\Theta$ for a single cell is $1/2$. Substitution of this into Eq. (2.34) gives Eq. (2.38) for the effective diffusion coefficient of the two cells. Therefore, the effective diffusion coefficient of the single cell is found to be identical to that of the two cells in contact. Similarly, the effective diffusion coefficients for the number of cells, corresponding to $2^2, 2^3 \ldots 2^n$ in contact are the same as that of the single unit cell. The effective diffusion coefficient or diffusional dispersion term is thus expected to be the same as that of the unit cell.

The transient time which is proportional to the square of the particle size will be longer for larger particles; but when fluid is flowing, the effective diffusion coefficient is considered to attain a steady-state value in a much shorter time.

Anyway, the transient time is usually so short compared to the mean residence time that the dispersion coefficient for an adsorbing species may be assumed to be constant over the entire adsorption period. Moreover, the dispersion coefficient for an adsorbing species is generally considered to be equal to that under inert conditions. In fact, Kaguei *et al.* [22] measured, in packed beds of activated carbon particles, the dispersion coefficients for adsorbing species, nitrogen, imposed on a carrier stream of hydrogen. They found that the dispersion coefficients under adsorption conditions were practically the same as those for an inert bed.

REFERENCES

[1]   P. V. Danckwerts, *Chem. Eng. Sci.* **2**, 1 (1953).
[2]   J. F. Wehner and R. H. Wilhelm, *Chem. Eng. Sci.* **6**, 89 (1956).
[3]   E. T. van der Laan, *Chem. Eng. Sci.* **7**, 187 (1958).
[4]   L. T. Fan and R. C. Bailie, *Chem. Eng. Sci.* **13**, 63 (1960).
[5]   K. B. Bischoff, *Chem. Eng. Sci.* **16**, 131 (1961).
[6]   K. B. Bischoff and O. Levenspiel, *Chem. Eng. Sci.* **17**, 245 (1962).
[7]   L. T. Fan and Y. K. Ahn, *Ind. Eng. Chem. Process Des. Dev.* **1**, 190 (1962).
[8]   J. J. Carberry and M. M. Wendel, *AIChE J.* **9**, 129 (1963).
[9]   S. L. Liu and N. R. Amundson, *Ind. Eng. Chem. Fund.* **2**, 183 (1963).
[10]   H. Hofmann and H. J. Astheimer, *Chem. Eng. Sci.* **18**, 643 (1963).
[11]   W. R. Schmeal and N. R. Amundson, *AIChE J.* **12**, 1202 (1966).
[12]   D. E. Mears, *Chem. Eng. Sci.* **26**, 1361 (1971).
[13]   C. Y. Wen and L. T. Fan, *Models for Flow Systems and Chemical Reactors*, Marcel Dekker, New York (1975).
[14]   K. S. Chang, K. Bergevin and E. W. Godsave, *J. Chem. Eng. Japan* **15**, 126 (1982).
[15]   D. J. Gunn, *Trans. Inst. Chem. Eng.* **46**, CE153 (1968).
[16]   R. G. Carbonell, *Chem. Eng. Sci.* **34**, 1031 (1979).

[17]   N. Wakao, S. Kaguei and H. Nagai, *Chem. Eng. Sci.* **33**, 183 (1978).
[18]   M. F. Edwards and J. F. Richardson, *Chem. Eng. Sci.* **23**, 109 (1968).
[19]   E. V. Evans and C. N. Kenney, *Trans. Inst. Chem. Eng.* **44**, T189 (1966).
[20]   M. Suzuki and J. M. Smith, *Chem. Eng. J.* **3**, 256 (1972).
[21]   N. Wakao, Y. Iida and S. Tanisho, *J. Chem. Eng. Japan* **7**, 438 (1974).
[22]   S. Kaguei, D. I. Lee and N. Wakao, *Kagaku Kogaku Ronbunshu* **6**, 397 (1980).

3 Diffusion and Reaction in a Porous Catalyst

Porous solid catalysts used for gas catalytic reactions have specific surface areas of tens to hundreds of square meters per gram. This enormous amount of surface area results mainly from the fine interconnecting pores in the catalyst pellet.† If a chemical reaction is very fast, it proceeds only at the external surface of the pellet. If, however, the reaction is very slow, the reactant gas may diffuse deep into the pores of pellet, even to the center of the pellet, and the chemical reaction takes place everywhere uniformly in the pellet.

In the laboratory, reaction rate determined directly by measurements using a differential reactor, for example, is the overall rate. The overall rate constant does not necessarily mean the intrinsic chemical reaction rate constant. For the design of industrial packed bed reactors, one needs to know the overall reaction rate, not the intrinsic chemical reaction rate. The overall rate is governed not only by the chemical reaction, but also by the diffusion rate through the pores inside the catalyst pellet as well as at the pellet's external surface. If we simply measure activation energy from the overall reaction rate constants, the activation energy may differ from that of the intrinsic chemical reaction. The importance of diffusion is often underestimated by some catalyst chemists.

Wheeler [1,2] made an extensive study on the role of pore diffusion in catalysis. Also, Dullien [3], Jackson [4], Petersen [5], Satterfield [6] and Smith [7] have reviewed the subject of pore diffusion associated with chemical reaction well. The review article of Youngquist [8] will also help readers understand the basic principles of diffusion and reaction in a porous catalyst.

In this chapter, the validity of the assumption of intrapellet concentration being radially symmetric is examined first. The importance of the

---

† A catalyst pellet is often made by compressing fine particles. In this chapter the word "pellet" is used to distinguish it from the fine particles.

94

catalyst effectiveness factor and the Jüttner modulus in the catalytic reaction system, their relationships and evaluation for various systems are then discussed. In the section on pore diffusion of gases, the mechanisms of pore diffusion and their interpretations are elucidated; moreover, the measurement of effective diffusivities and their predictions using various proposed models are reviewed.

3.1   Assumption of a Concentric Concentration Profile in a Spherical Catalyst Pellet

The objective of this section is to examine whether the steady-state concentration profile in a catalyst pellet may be assumed to be radially symmetric when evaluating the reaction rate in a pellet.

Let us illustrate this by considering an isothermal, first-order, irreversible reaction proceeding in a spherical catalyst pellet of radius, R, under steady-state conditions. The mass balance equation for a reactant is given by Eq. (2.16) in Section 2.2.

Integration of Eq. (2.16) over a surface area with radius r and dividing this by $4\pi r^2$ gives

$$\frac{1}{4\pi r^2} \int_0^{2\pi} d\Phi \int_{-1}^1 (D_e \nabla^2 c^* - k_x c^*) r^2 \, d\cos\theta = 0. \tag{3.1}$$

Following the same procedure as given in Section 1.2.4, Eq. (3.1) becomes, under isothermal conditions:

$$D_e \frac{1}{r^2} \frac{d}{dr}\left(r^2 \frac{dc}{dr}\right) - k_x c = 0 \tag{3.2}$$

where

$$c = \frac{1}{4\pi} \int_0^{2\pi} d\Phi \int_{-1}^1 c^* \, d\cos\theta. \tag{3.2a}$$

This c represents an average concentration over a surface with radius r in the pellet; therefore, the assumption of radially symmetric intrapellet concentration may be used in evaluating the chemical reaction rate if the reaction is first-order with respect to the reactant concentration.

This assumption can also be verified by considering the fact that the solution to Eq. (2.16) may be expressed by a series of Legendre functions:

$$c^*(r, \theta, \Phi) = \sum_{n=0}^{\infty} \sum_{m=-n}^{n} f_n^m(r) P_n^m(\cos\theta) \exp(im\Phi) \qquad (3.3)$$

where $P_n^m(\cos\theta)$, with m defined in the range $-n$ to n, is an associate Legendre function of the first kind. The coefficient, $f_n^m(r)$, is given by

$$f_n^m(r) = (-1)^m \left(\frac{2n+1}{4\pi}\right) \int_0^{2\pi} d\Phi \int_{-1}^{1} c^*(r, \theta, \Phi) P_n^{-m}(\cos\theta)$$

$$\times \exp(-im\Phi) \, d\cos\theta \qquad (3.4)$$

which should satisfy Eq. (3.5).

$$D_e \left[\frac{1}{r^2} \frac{d}{dr} \left(r^2 \frac{df_n^m}{dr}\right) - \frac{n(n+1)}{r^2} f_n^m \right] - k_x f_n^m = 0. \qquad (3.5)$$

The overall reaction rate, R_p, throughout a pellet is:

$$R_p = -k_x \int c^* \, dv$$

$$= -k_x \int_0^{R} r^2 \, dr \int_0^{2\pi} d\Phi \int_{-1}^{1} c^*(r, \theta, \Phi) \, d\cos\theta. \qquad (3.6)$$

Substitution of Eq. (3.3) into (3.6) and consideration of the orthogonality of the Legendre function give

$$\int_0^{2\pi} d\Phi \int_{-1}^{1} P_n^m(\cos\theta) \exp(im\Phi) \, d\cos\theta = 4\pi \quad \text{when } n = 0$$
$$= 0 \quad \text{when } n \neq 0. \qquad (3.7)$$

The overall reaction rate is then

$$R_p = -4\pi k_x \int_0^R f_0^o(r)\, r^2 \, dr.$$

(3.8)

Since $P_0^o(\cos\theta) = 1$, Eq. (3.4) shows that

$$f_0^o(r) = \frac{1}{4\pi} \int_0^{2\pi} d\Phi \int_{-1}^1 c^*(r,\theta,\Phi)\, d\cos\theta.$$

(3.9)

Also, Eq. (3.5) reduces to

$$D_e \frac{1}{r^2} \frac{d}{dr}\left(r^2 \frac{df_0^o}{dr}\right) - k_x f_0^o = 0.$$

(3.10)

According to Eq. (3.8), the overall reaction rate in a pellet may be evaluated only in terms of $f_0^o(r)$. As Eq. (3.9) shows, this is an average concentration of c^* over a surface area with radius r. Therefore, although c^* was originally a function of r, θ and Φ, Eq. (3.10) indicates that $f_0^o(r)$ has center symmetry.

3.2 Effectiveness Factors for First-order Irreversible Reactions under Isothermal Conditions

A catalyst effectiveness factor is an indication of how much internal surface area is being utilized in a given reaction. The effectiveness factor depends not only upon the intrinsic chemical reaction rate, but also on the rates of the diffusion processes. If the reaction proceeds only at the external surface, the catalyst effectiveness factor is low, whereas if the internal pores are being used effectively for the chemical reaction, the factor is large.

The concept of a catalyst effectiveness factor was first introduced by the German scientist Jüttner [9] as early as 1909. The work of Thiele [10] on the catalyst effectiveness factor, published in 1939 and written in English, is popular among researchers in catalysis. Also in the late 1930's, a similar theoretical work was reported in a German paper by Damköhler [11] and in a Russian paper by Zeldowitsch [12]. The parameter indicating the reaction-to-diffusion rate ratio is widely known as the Thiele modulus. The importance of this parameter was, in fact, first recognized by Jüttner. We shall, therefore, call it the Jüttner modulus in this book.

3.2.1   Effectiveness Factors for a Spherical Catalyst and Overall Reaction Rate Constants

Let us consider again a spherical catalyst pellet of radius R, as shown in Figure 3.1. The steady-state mass balance equation for a reacting species is

$$D_e \frac{1}{r^2} \frac{d}{dr}\left(r^2 \frac{dc}{dr}\right) + r_v = 0 \qquad (3.11)$$

where r_v is the chemical reaction rate, defined as the production rate per unit volume of catalyst pellet. For a reactant, the rate is negative.

In the preceding section it was shown that, for a first-order reaction, the intrapellet concentration may be assumed to be of radial symmetry. When the reaction is first-order and irreversible with respect to the reactant concentration, c, Eq. (3.11) becomes Eq. (3.2).

Based on the following boundary conditions:

$$\frac{dc}{dr} = 0 \qquad \text{at } r = 0 \qquad (3.11a)$$

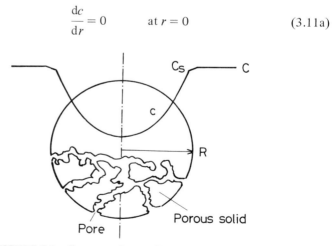

FIGURE 3.1   Concentration profile in a catalyst pellet.

and

$$c = C_s \qquad \text{at } r = R \qquad (3.11b)$$

where C_s is the reactant concentration at the surface of pellet, the solution to Eq. (3.2) under isothermal conditions is:

$$c = C_s \frac{R}{r} \frac{\sinh\left[r\left(\dfrac{k_x}{D_e}\right)^{1/2}\right]}{\sinh\left[R\left(\dfrac{k_x}{D_e}\right)^{1/2}\right]}. \qquad (3.12)$$

At steady-state conditions, the overall reaction rate throughout a single catalyst pellet is

$$R_p = 4\pi R^2 D_e \left(-\frac{dc}{dr}\right)_R \qquad (3.13)$$

or

$$R_p = -4\pi k_x \int_0^R cr^2 \, dr. \qquad (3.14)$$

Substitution of Eq. (3.12) into either Eq. (3.13) or (3.14) gives

$$R_p = -4\pi R D_e C_s \left\{ R\left(\frac{k_x}{D_e}\right)^{1/2} \coth\left[R\left(\frac{k_x}{D_e}\right)^{1/2}\right] - 1 \right\}. \qquad (3.15)$$

In terms of the Jüttner modulus, ϕ, which is defined as:

$$\phi = R\left(\frac{k_x}{D_e}\right)^{1/2} \qquad (3.16)$$

Eq. (3.15) becomes

$$R_p = -\frac{4\pi R^3}{3} k_x C_s \frac{3}{\phi}\left(\coth\phi - \frac{1}{\phi}\right). \qquad (3.17)$$

As a matter of fact, $(4\pi R^3/3)\,k_x C_s$ is the overall reaction rate in a pellet when the intrapellet concentration is C_s everywhere, in other words, there is no intrapellet diffusion resistance. Equation (3.17) is expressed as:

$$R_p = -\frac{4\pi R^3}{3}\,k_x C_s E_f \tag{3.18}$$

where

$$E_f = \frac{3}{\phi}\left(\coth\phi - \frac{1}{\phi}\right). \tag{3.19}$$

This E_f is the catalyst effectiveness factor, or sometimes called the catalyst utilization factor. The variation of E_f with ϕ, according to Eq. (3.19), for a first-order irreversible reaction is shown in Figure 3.2 over a wide range of ϕ. As shown, E_f is approximately given as:

In a chemical reaction controlling region

$$E_f = 1 \qquad \text{for } \phi < 0.6 \tag{3.20}$$

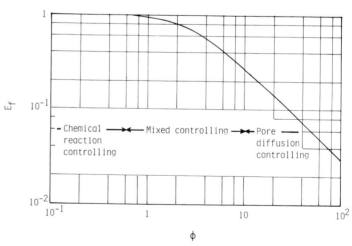

FIGURE 3.2   Catalyst effectiveness factor versus Jüttner modulus, for a first-order reaction in a spherical catalyst pellet.

In a pore diffusion controlling region

$$E_f = \frac{3}{\phi} \qquad \text{for } \phi > 10. \qquad (3.21)$$

A small Jüttner modulus means that the catalyst pellet is small, the chemical reaction is slow and/or there is a high intrapellet diffusion rate. Under these conditions, chemical reaction takes place uniformly through-out the pellet and thus the catalyst effectiveness factor is unity. On the other hand, if the catalyst pellet is large and chemical reaction is fast and/or intrapellet diffusion is slow, then, the Jüttner modulus is large, and consequently, the catalyst effectiveness factor becomes low.

Under steady-state conditions, the overall reaction rate is equal to the rate at which the reactants are supplied to the pellet surface from the bulk fluid. Therefore, Eq. (3.18) is related to the rate of mass transfer of the reacting species as follows:

$$-R_p = \frac{4\pi R^3}{3} k_x C_s E_f = 4\pi R^2 k_f (C - C_s)$$

$$= \frac{4\pi R^3}{3} \frac{C}{\dfrac{1}{k_x E_f} + \dfrac{R}{3k_f}} \qquad (3.22)$$

where C is the reactant concentration in the bulk fluid and k_f is the mass transfer coefficient at the pellet surface.

The reaction rate, r_x, defined on the basis of the unit volume of a packed bed catalytic reactor, is

$$r_x = \frac{1 - \epsilon_b}{\dfrac{4\pi R^3}{3}} R_p. \qquad (3.23)$$

The overall rate constant, K, of Eq. (2.2) is then

$$K = \frac{1 - \epsilon_b}{\dfrac{1}{k_x E_f} + \dfrac{R}{3k_f}}. \qquad (3.24)$$

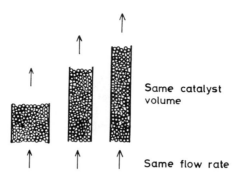

FIGURE 3.3   Reaction rate measurements with different sized columns.

Suppose, as illustrated in Figure 3.3, reaction rate measurements are to be made with several different sized cylindrical columns of packed bed reactors under the same reaction conditions (same volumetric flow rates, catalyst volume, temperature and pressure). Everything is the same except for the column size or the fluid velocity. In Eq. (3.24), the intrapellet term, $k_x E_f$, is not affected by flow rate, but the mass transfer coefficient, k_f, depends upon the fluid velocity. The mass transfer coefficient is low at low fluid velocity, but increases with an increase in fluid velocity (see Chapter 4). Therefore, the overall rate constants and the conversion are low at low fluid velocities in large columns, and high at high fluid velocities in small columns. However, if the size of the column is sufficiently small, or the fluid velocity is high enough, such that, the external diffusional resistance is negligible, then the overall rate constant becomes

$$K = (1 - \epsilon_b) k_x E_f. \tag{3.25}$$

Under such conditions, the overall rate constant and the conversion do not depend on the fluid velocity any more. This is the region in which the intrapellet diffusion/reaction is rate controlling. This may be further classified into chemical reaction control and/or pore diffusion control, as governed by the following equations:

In a chemical reaction controlling region

$$K = (1 - \epsilon_b) k_x \qquad \text{for } \phi < 0.6 \tag{3.26}$$

In a pore diffusion controlling region

$$K = \frac{3(1 - \epsilon_b)}{R} (D_e k_x)^{1/2} \qquad \text{for } \phi > 10. \qquad (3.27)$$

The region, where the overall rate constant depends upon the fluid velocity, is extrapellet diffusion controlling. The overall rate constant is:

In an extrapellet diffusion controlling region

$$K = \frac{3(1 - \epsilon_b)}{R} k_f. \qquad (3.28)$$

It is also interesting to note that the overall rate constants are inversely proportional to the catalyst pellet size when diffusion, either externally or internally, is the rate controlling step.

The intrinsic chemical reaction rate constant, k_x, is expressed in terms of the activation energy, E, as:

$$k_x = k_0 \exp\left(-\frac{E}{R_g T}\right) \qquad (3.29)$$

where R_g is the gas constant and T is temperature.

The overall rate constant, K, is also expressed as:

$$K = K_0 \exp\left(-\frac{E'}{R_g T}\right) \qquad (3.30)$$

where E' is the activation energy based on the overall rate constant. Therefore, from Eqs. (3.25), (3.29) and (3.30), we obtain

$$\ln K_0 - \frac{E'}{R_g T} = \ln (1 - \epsilon_b) + \ln k_0 - \frac{E}{R_g T} + \ln E_f. \qquad (3.30a)$$

Differentiation of this with respect to $-1/(R_g T)$ gives

$$E' = E + \frac{d \ln E_f}{d\left(-\dfrac{1}{R_g T}\right)}$$

$$= E + \frac{d\ln E_f}{d\ln \phi} \frac{d\ln \phi}{d\ln k_x} \frac{d\ln k_x}{d\left(-\dfrac{1}{R_g T}\right)}. \tag{3.31}$$

With Eqs. (3.16) and (3.29), we obtain

$$E' = E\left(1 + \frac{1}{2}\frac{d\ln E_f}{d\ln \phi}\right). \tag{3.32}$$

From Eqs. (3.20) and (3.21), Eq. (3.32) is simplified to:

$$\left.\begin{aligned} E' &= E & \text{for } \phi < 0.6 \\ E' &= \frac{E}{2} & \text{for } \phi > 10. \end{aligned}\right\} \tag{3.32a}$$

If the overall rate constants are measured in a pore diffusion controlling region, an Arrhenius plot of the rate constants is expected to give half the activation energy of the intrinsic chemical reaction. This is illustrated in Figure 3.4. At high temperatures, $k_x E_f \gg 3k_f/R$ and the reaction rate is extrapellet diffusion controlling. In this region, an Arrhenius plot of K will yield a small activation energy, of the order of about 10 kJ mol^{-1}, corresponding to the value of k_f.

3.2.2   Effectiveness Factors for Other Geometries

Effectiveness factors for non-spherical pellets in which a first-order chemical reaction is occurring under isothermal and steady-state conditions are listed below:

3.2.2.1   *Flat plate (of height L) with one side and edges sealed*

If the plate is contacted with reactants on one side only, then:

$$E_f = \frac{\tanh \phi_L}{\phi_L} \tag{3.33}$$

where $\phi_L = L(k_x/D_e)^{1/2}$.

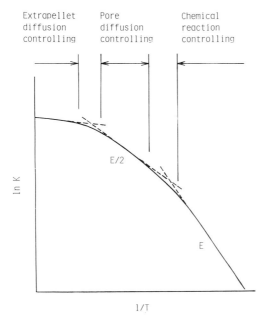

FIGURE 3.4   Arrhenius plot of overall rate constants.

3.2.2.2   Flat plate (of height 2L) with edges sealed

If the plate is contacted with reactants on both sides, then:

$$E_f = \text{Eq. (3.33).}$$

3.2.2.3   Infinitely long cylinder (of radius R) or finite length cylinder (of radius R) with ends sealed

If the cylinder is contacted with reactants on the cylinder surface [13], then:

$$E_f = \frac{2 I_1(\phi)}{\phi \, I_0(\phi)} \tag{3.34}$$

where $\phi = R(k_x/D_e)^{1/2}$; I_0 and I_1 are the modified Bessel functions of the first kind, respectively, of zeroth and first-order.

3.2.2.4 *Finite length cylinder (of radius R and height 2L)*

If the cylinder is contacted with reactants on all surfaces [14], then:

$$E_f = 1 - 8 \sum_{m=1}^{\infty} \sum_{n=1}^{\infty} \frac{1}{j_m^2 k_n^2} \left[\frac{\phi^2}{\phi^2 + j_m^2 + (k_n \rho)^2} \right] \qquad (3.35)$$

where

$$k_n = \frac{(2n-1)\pi}{2}$$

$$\rho = \frac{R}{L}$$

$$\phi = R \left(\frac{k_x}{D_e} \right)^{1/2}$$

and j_m is an m-th root of the following Bessel function of the first kind and zeroth-order:

$$J_0(j_m) = 0.$$

3.2.2.5 *Ring (of inside radius R_i, outside radius R and height 2L)*

If the ring is contacted with reactants on all surfaces [14], then:

$$E_f = 1 - \frac{8}{1 - \beta^2} \sum_{m=1}^{\infty} \sum_{n=1}^{\infty} \frac{1}{j_m^2 k_n^2} \left[\frac{J_0(\beta j_m) - J_0(j_m)}{J_0(\beta j_m) + J_0(j_m)} \right] \left[\frac{\phi^2}{\phi^2 + j_m^2 + (k_n \rho)^2} \right]$$

$$(3.36)$$

where

$$k_n = \frac{(2n-1)\pi}{2}$$

$$\rho = \frac{R}{L}$$

$$\phi = R \left(\frac{k_x}{D_e} \right)^{1/2}$$

$$\beta = \frac{R_i}{R}$$

and j_m is an m-th root of

$$J_0(\beta j_m) \, Y_0(j_m) - J_0(j_m) \, Y_0(\beta j_m) = 0$$

where Y_0 is the Bessel function of the second kind and zeroth-order.

For a ring (with sealed ends) contacted with reactants on the inside and outside surfaces [14], then:

$$E_f = \frac{2}{(1-\beta^2) \, \phi} + \frac{\begin{array}{l} \{ [K_0(\beta\phi) - K_0(\phi)] \, [I_1(\phi) - \beta I_1(\beta\phi)] \\ + [I_0(\phi) - I_0(\beta\phi)] \, [\beta K_1(\beta\phi) - K_1(\phi)] \} \end{array}}{K_0(\beta\phi) \, I_0(\phi) - I_0(\beta\phi) \, K_0(\phi)} \quad (3.37)$$

where K_0 and K_1 are modified Bessel functions of the second kind of zeroth and first-order, respectively.

For a ring (with sealed ends and inside surface) contacted with reactants on the outside surface [14], then:

$$E_f = \frac{2}{(1-\beta^2) \, \phi} \left[\frac{I_1(\phi) \, K_1(\beta\phi) - I_1(\beta\phi) \, K_1(\phi)}{I_0(\phi) \, K_1(\beta\phi) + I_1(\beta\phi) \, K_0(\phi)} \right]. \quad (3.38)$$

For a ring (with sealed ends and outside surface) contacted with reactants on the inside surface [14], then:

$$E_f = \frac{2\beta}{(1-\beta^2) \, \phi} \left[\frac{I_1(\phi) \, K_1(\beta\phi) - I_1(\beta\phi) \, K_1(\phi)}{I_1(\phi) \, K_0(\beta\phi) + I_0(\beta\phi) \, K_1(\phi)} \right]. \quad (3.39)$$

3.2.2.6 *Sphere (of radius R) with the catalyst coated on an inert solid sphere (of radius R_i)* [14]

$$E_f = \frac{3}{(1-\beta^3) \, \phi} \left[\frac{\coth[(1-\beta) \, \phi] + \beta\phi}{1 + \beta\phi \coth[(1-\beta) \, \phi]} - \frac{1}{\phi} \right] \quad (3.40)$$

where ϕ and β are the same as those defined in Section 3.2.2.5.

The effectiveness factors for extruded catalysts with cross-sections, such as a letter L, a dumb-bell, a trilobe or a quadrulobe, have also been numerically computed by Suzuki and Uchida [15]. Low effectiveness factor means that the reaction takes place in the vicinity of surfaces exposed to the reactants. Aris [13] has shown that Eq. (3.33) holds for any shape of catalyst pellet at high values of ϕ, if L is taken as the ratio of the pellet volume to the area of surfaces exposed to the reactants.

3.3   Pore Diffusion of Gases

In porous media, diffusion of non-adsorbing inert gas occurs through the pore volumes. There are two types of pore volume diffusion. If the pore is large, normal diffusion due to molecule–molecule collisions takes place. If the pore diameter is smaller than the length of the mean free path of the gas molecules, diffusion proceeds by molecule–wall collisions, and is called Knudsen diffusion [16]. If the pores are filled with liquid, bulk liquid phase diffusion is the only transport process.

For physical adsorbing species, diffusion takes place not only through the pore volumes, but also along the pore surfaces. The physically adsorbed molecules migrate on to the pore walls. The amount adsorbed is in equilibrium with its gas phase concentration, and consequently, pore volume diffusion and surface diffusion both proceed in the direction of decreasing gas phase concentration in the pore.

At low temperatures, the mechanism of adsorption is predominantly physical. Under such conditions, overall intrapellet diffusion is largely due to surface diffusion. For catalytic reaction at temperatures higher than the boiling points of the reacting species, surface diffusion is generally considered to be of little importance [1, 2].

3.3.1   Diffusion in a Capillary Tube

3.3.1.1   *Knudsen diffusion*

Consider the steady-state countercurrent diffusion of two inert gases, 1 and 2, in a capillary tube (radius a and length L) in which the pressures at both ends are kept constant. If the diameter of the capillary tube is smaller than the length of the mean free path of the species, diffusion takes place only by molecule–wall collisions. There is little chance of molecule–molecule collisions occurring in the capillary tube, and molecules collide only with the capillary tube wall. If the concentration of the species under consideration is higher at one end, say the inlet end ($x = 0$), than the other, outlet end ($x = L$), molecules collide with the tube wall most frequently at the inlet end. The frequency of collision decreases with increasing distance of x, and then the species diffuse in the capillary tube from the inlet toward the outlet.

In the Knudsen region, the diffusion flux is proportional to the difference in concentration between the ends of the capillary tube, but inversely proportional to the tube length. The molar flux for species 1, J_1, across a distance, L, in the direction of increasing x is

$$J_1 = D_{K1} \frac{\Delta C_1}{L} \tag{3.41}$$

where $\Delta C_1 = (C_1)_{x=0} - (C_1)_{x=L}$ and D_{K1} is the Knudsen diffusivity of species 1.

In general, Knudsen diffusivity, D_K, is related to the mean molecular velocity, $\bar{v}$, and the capillary tube radius, a, by the following equation:

$$D_K = \frac{2\bar{v}a}{3}. \tag{3.42}$$

For an ideal gas of molar mass, M, the velocity is

$$\bar{v} = \left(\frac{8R_gT}{\pi M}\right)^{1/2}. \tag{3.43}$$

Thus

$$D_K = \frac{2a}{3}\left(\frac{8R_gT}{\pi M}\right)^{1/2}. \tag{3.44}$$

In SI units, the above equation is given as:

$$D_K = 3.068a\left(\frac{T}{M}\right)^{1/2} \tag{3.44a}$$

where D_K is in m^2 s^{-1}, a is in m, T is in K and M is in kg mol^{-1}.

Similarly, the flux of species 2 in the Knudsen diffusion region is

$$J_2 = D_{K2} \frac{\Delta C_2}{L}. \tag{3.45}$$

Equations (3.41) and (3.45) are always valid even if the total pressures at either end of the capillary tube are not equal. However, if the ends are at the same pressure, the diffusion flux ratio becomes

$$\frac{J_1}{J_2} = -\left(\frac{M_2}{M_1}\right)^{1/2}. \tag{3.46}$$

In the Knudsen diffusion region, the molecules collide only with the capillary tube wall, so that there is no interaction between J_1 and J_2. However, if the capillary tube is large, the situation is completely different. Molecules of species 1 collide with molecules of species 2 as well as with its own kind. This is the normal or bulk diffusion.

3.3.1.2   Normal diffusion

In the normal diffusion region, the flux is expressed by

$$J_1 = -D_{12} \frac{dc_1}{dx} + y_1(J_1 + J_2) \tag{3.47}$$

where c_1 is the concentration of species 1 in a capillary tube, y_1 is the mole fraction of species 1, and D_{12} is the binary diffusion coefficient for species 1 and 2.

The first term on the right hand side of Eq. (3.47) is Fick's first law of diffusion. In the second term, $J_1 + J_2$ is the total molar flux or diffusive flow of mixture of species 1 and 2 in the direction of increasing x, and $y_1(J_1 + J_2)$ is the transport rate of species 1 by the flow. The diffusion flux of species 1, relative to a fixed coordinate system, is the sum of the Fick's diffusion flux and the flux carried by the diffusive flow.

Similarly, the diffusion flux for species 2 is

$$J_2 = -D_{21} \frac{dc_2}{dx} + y_2(J_1 + J_2). \tag{3.48}$$

Note that

$$D_{12} = D_{21}. \tag{3.49}$$

3.3.1.3 Combined Knudsen diffusion and normal diffusion

If both Knudsen and normal diffusion occur simultaneously then the diffusion flux may be expressed by the following form:

$$J_1 = -\left(\frac{1}{\dfrac{1}{D_{12}} + \dfrac{1}{D_{K1}}}\right)\frac{dc_1}{dx} + \frac{y_1(J_1 + J_2)}{1 + \dfrac{D_{12}}{D_{K1}}} \tag{3.50}$$

or

$$J_1 = -\left(\frac{1}{\dfrac{1 - \alpha y_1}{D_{12}} + \dfrac{1}{D_{K1}}}\right)\frac{dc_1}{dx} \tag{3.50a}$$

where

$$\alpha = 1 + \frac{J_2}{J_1}. \tag{3.50b}$$

Equation (3.50) reduces to Eq. (3.47) for normal diffusion when the capillary tube is large, or $D_{K1} \gg D_{12}$.

When $D_{K1} \ll D_{12}$, or the capillary tube is small and diffusion is of the Knudsen type, Eq. (3.50) becomes

$$J_1 = -D_{K1}\frac{dc_1}{dx}. \tag{3.51}$$

Equation (3.41) is an integrated form of Eq. (3.51) over a distance L.

Equation (3.50) was derived in 1961 by Evans *et al.* [17] to describe the diffusion processes from normal to Knudsen diffusion through a transition region. In the diffusion equations based on Chapman–Enskog kinetic theory for a multicomponent system, Evans *et al.* assumed that one of the components consisted of giant gas molecules which were uniformly distributed and fixed in space. Thus, from the dusty gas model, they obtained Eq. (3.50) for a binary gas system.

Scott and Dullien [18] obtained Eq. (3.50) from momentum transfer arguments. Rothfeld [19] also derived, from similar momentum transfer discussions, a diffusion equation for gas in porous media, which reduces to Eq. (3.50) in the case of diffusion in a single capillary tube.

Rearranging Eq. (3.50) for species 1 gives

$$\left(1 + \frac{D_{12}}{D_{K1}}\right) J_1 = -D_{12}\frac{dc_1}{dx} + y_1(J_1 + J_2). \tag{3.52}$$

Similarly, for species 2

$$\left(1 + \frac{D_{21}}{D_{K2}}\right) J_2 = -D_{21}\frac{dc_2}{dx} + y_2(J_1 + J_2). \tag{3.53}$$

Since $D_{12} = D_{21}$, addition of Eqs. (3.52) and (3.53), at constant total pressure, gives

$$\frac{J_1}{D_{K1}} + \frac{J_2}{D_{K2}} = 0 \tag{3.54}$$

which further reduces to Eq. (3.46). The ratio of constant pressure diffusion fluxes is, thus, always governed by the relationship in Eq. (3.46), irrespective of the modes of diffusion: Knudsen, transition or normal.

3.3.1.4   Self-diffusion in a capillary tube

Self-diffusion occurs in a single species system. The flux of self-diffusion is usually measured by the countercurrent diffusion of isotopes technique. If J_1 is the flux of the isotopes or labeled molecules, and J_2 is the flux of the unlabeled molecules; since $M_1 = M_2$, Eq. (3.46) shows that $J_1 + J_2 = 0$. The flux of self-diffusion in a capillary tube is then:

$$J_1 = -D_1\frac{dc_1}{dx} \tag{3.55}$$

where

$$\frac{1}{D_1} = \frac{1}{D_{11}} + \frac{1}{D_{K1}}. \tag{3.55a}$$

As a matter of fact, Bosanquet [20] obtained Eq. (3.55a) by assuming that the resistance to transport is the sum of the resistances due to both molecule-wall collisions and self-diffusion. In Eq. (3.55a), D_{11} is the self-diffusion coefficient defined as:

$$D_{11} = \frac{\bar{v}_1 \lambda_1}{3} \tag{3.56}$$

where λ_1 is the length of the mean free path of the molecules of species 1.

Wheeler [1, 2] developed an exponential combining formula for D_1:

$$D_1 = D_{11}\left[1 - \exp\left(-\frac{D_{K1}}{D_{11}}\right)\right] \tag{3.57}$$

where, from Eqs. (3.42) and (3.56), it is shown that

$$\frac{D_{K1}}{D_{11}} = \frac{2a}{\lambda_1}. \tag{3.57a}$$

Pollard and Present [21] carried out elaborate calculations for D_1 from molecular theory; the result is

$$D_1 = D_{11}\left(1 - \frac{3}{8}\frac{\lambda_1}{a} + \frac{6}{\pi}\frac{\lambda_1}{a}Q\right) \tag{3.58}$$

where

$$Q = \frac{\pi}{16} - \frac{\pi}{6}\frac{a}{\lambda_1} + \frac{\pi}{3}\left(\frac{a}{\lambda_1}\right)^2 - \frac{\pi}{3}\left(1.2264 - \frac{3}{4}\ln\frac{2\gamma a}{\lambda_1}\right)\left(\frac{a}{\lambda_1}\right)^3 + \dots \tag{3.58a}$$

in which γ is Euler's constant. As shown in Figure 3.5, the Bosanquet [20]

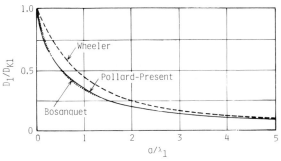

FIGURE 3.5   Comparison of self-diffusion coefficient formulae.

formula obtained by the additive resistance law is in surprisingly good agreement with the rigorous expression of Pollard and Present.

3.3.2   Diffusion in a Porous Solid

Effective diffusivity, D_e, in a porous solid is defined by the following equation:

$$N = AD_e \frac{\Delta C}{L} \qquad (3.59)$$

where N is the diffusion rate, A is the cross-sectional area perpendicular to the direction of diffusion and ΔC is the difference in concentration across the distance L.

3.3.2.1   *Measurement of effective diffusivity in a porous solid*

The Wicke–Kallenbach [22, 23] type of apparatus shown in Figure 3.6 is widely used for effective diffusivity measurements in a porous solid. When gas 1 and gas 2 sweep across the top and bottom of the cylinder of a porous solid under constant total pressure, countercurrent diffusion takes place through the pores of the cylinder. From measurements of concentration and flow rate of exit streams under steady-state conditions, the steady countercurrent diffusion rate is determined. The effective diffusivity defined by Eq. (3.59) is then evaluated.

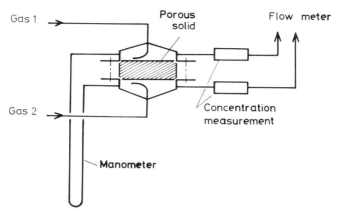

FIGURE 3.6   Wicke–Kallenbach type of apparatus for steady countercurrent
diffusion measurements.

For a cylindrical pellet, the cross-sectional area is a constant, and the
concentration profiles of the diffusing species in the pellet are linear and
parallel under steady-state conditions. For spherical particles, however,
the cross-sectional area varies, and as the calculations made by Kaguei
et al. [24] show: the steady-state concentration profiles are nonlinear, as
shown in Figure 3.7. This makes it rather difficult to define A in Eq.
(3.59). Kaguei *et al*. [24] devised a method for the evaluation of the
effective diffusivity in spherical pellets. In their method, they suggested
that the pellets should be glued into the holes made through a plate with a
thickness less than the diameter of the pellet. The ends of the pellets
sticking out of the plate should then be shaved off, as shown in Figure
3.8. They recommended that the average cross-sectional area, which
should be taken as A in Eq. (3.59), should be found from Figure 3.9.
In the graph A_1 and A_2 are the cross-sectional areas of the pellet exposed
on either side of the plate. Also, A_0 is the cross-sectional area of the
spherical pellet at the equator, or $A_0 = \pi R^2$, where R is the radius of the
pellet.

In the range where both A_1/A_0 and A_2/A_0 are greater than 0.3, the
curves shown in Figure 3.9 are expressed within an error of 1% by the use
of log-mean areas, A_1' and A_2', defined as follows:

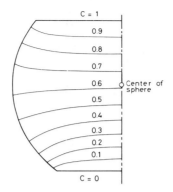

FIGURE 3.7   Steady-state concentration profiles in spherical pellet with sides
shaved off.

$$A = \frac{L}{\dfrac{L_1}{A_1'} + \dfrac{L_2}{A_2'}} \qquad (3.60)$$

where

$$A_1' = \frac{A_0 - A_1}{\ln\left(\dfrac{A_0}{A_1}\right)}$$

$$\text{(3.60a)}$$

$$A_2' = \frac{A_0 - A_2}{\ln\left(\dfrac{A_0}{A_2}\right)}$$

and L_1 and L_2 are the distances between the center of the sphere and the plate surfaces (refer to Figure 3.8); $L = L_1 + L_2$. With the mean area, A, the effective diffusivity, D_e, for a spherical pellet is determined from measurements of the countercurrent diffusion rate, N, and the difference in concentration, ΔC, across the distance L.

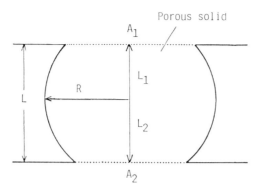

FIGURE 3.8   Sphere with sides shaved off.

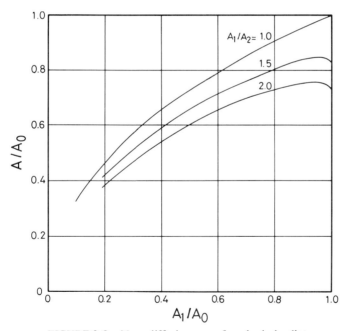

FIGURE 3.9   Mean diffusion area of a spherical pellet.

3.3.2.2   *Prediction of effective diffusivity from the proposed models*

Wheeler's [1, 2] parallel pore model assumes that the porous structure can be represented by a number of parallel capillaries of the same size. If there are n such capillary tubes of radius, $\bar{a}$, and length, L_e, in a unit mass of porous solid of height, L, the total surface area of the tubes is $S_g = n2\pi\bar{a}L_e$, and the total tube volume is $V_g = n\pi(\bar{a})^2 L_e$. The model further assumes that S_g is equal to the internal surface area usually measured in a BET apparatus, and V_g is equal to the pore volume of the solid. The mean pore radius is then calculated by

$$\bar{a} = \frac{2V_g}{S_g}. \tag{3.61}$$

The diffusion flux in a pore of mean radius, $\bar{a}$, is given by Eq. (3.62).

$$J_1 = -\left(\frac{1}{\dfrac{1-\alpha y_1}{D_{12}} + \dfrac{1}{\bar{D}_{K1}}}\right) \frac{dc_1}{dx} \tag{3.62}$$

where

$$\bar{D}_{K1} = \frac{2\bar{v}_1\bar{a}}{3}. \tag{3.62a}$$

The diffusion rate per unit solid area, J_e, is related to the diffusion flux, J, through a straight capillary tube of radius $\bar{a}$ and length L by:

$$J_e = \frac{\epsilon_p}{\tau} J \tag{3.63}$$

$$= \delta J \tag{3.63a}$$

where τ and δ are the tortuosity factor and diffusibility, respectively.

The tortuosity factor and diffusibility have been studied by many investigators. Wheeler [2] suggests that the diffusion rate per unit solid area is

$$J_e = m\pi(\bar{a})^2 \frac{L}{L_e} J \tag{3.64}$$

where m is the number of pores per unit solid area. The porosity of the solid, ϵ_p, is equal to $m\pi(\bar{a})^2 L_e/L$, so that Eq. (3.64) becomes

$$J_e = \epsilon_p \left(\frac{L}{L_e}\right)^2 J. \tag{3.65}$$

Moreover, Wheeler assumes that the pores intersect any plane at an angle of $\pi/4$ rad, or $L_e/L = 2^{1/2}$. The tortuosity factor of Eq. (3.63) is then 2. On the other hand, the random pore model of Wakao and Smith [25], which will be described later, predicts

$$\delta = \epsilon_p^2. \tag{3.66}$$

The diffusion flux is expressed in terms of effective diffusivity as:

$$J_{e1} = -D_{e1} \frac{dc_1}{dx} \tag{3.67}$$

where

$$D_{e1} = \delta \frac{1}{\dfrac{1 - \alpha y_1}{D_{12}} + \dfrac{1}{\bar{D}_{K1}}}. \tag{3.67a}$$

The diffusion flux measured in a Wicke-Kallenbach apparatus corresponds to the one obtained from the integration of Eq. (3.67):

$$J_{e1} = \delta \frac{D_{12}P}{\alpha R_g TL} \ln \left[\frac{1 + \dfrac{D_{12}}{\bar{D}_{K1}} - \alpha(y_1)_{out}}{1 + \dfrac{D_{12}}{\bar{D}_{K1}} - \alpha(y_1)_{in}} \right] \tag{3.68}$$

where P is the total pressure. The effective diffusivity is then

$$D_{e1} = \delta \frac{D_{12}}{\alpha[(y_1)_{in} - (y_1)_{out}]} \ln \left[\frac{1 + \dfrac{D_{12}}{\bar{D}_{K1}} - \alpha(y_1)_{out}}{1 + \dfrac{D_{12}}{\bar{D}_{K1}} - \alpha(y_1)_{in}} \right]. \tag{3.69}$$

Similar to Eq. (3.46), the diffusion fluxes through a porous solid at uniform total pressure are related by Eq. (3.70).

$$\frac{J_{e1}}{J_{e2}} = -\left(\frac{M_2}{M_1}\right)^{1/2}. \qquad (3.70)$$

In the earlier work, Eq. (3.70) was considered to hold only in the Knudsen region, whereas in the normal diffusion region the assumption of equimolar countercurrent diffusion according to Eq. (3.71) had often been used.

$$J_{e1} + J_{e2} = 0. \qquad (3.71)$$

This invalid assumption could have arisen due to confusion with diffusion in a closed vessel.

Hoogschagen [26], Henry et al. [27], Evans et al. [28] and Scott and Cox [29] measured both fluxes in steady countercurrent diffusion. They all observed that the measured fluxes were inversely proportional to the square roots of the molecular weights. Hoogschagen showed that the pore sizes of the tablet used in their experiments were much larger than the length of the mean free path of the gas molecules. Evans et al. and Scott and Cox also showed that their data were obtained within the normal diffusion region. Based on the assumption that normal diffusion would be equimolar, Henry et al. regarded their experimental data as proof of Knudsen diffusion.

For porous solids having monodisperse pores, the diffusion flux may be estimated by Eq. (3.63). However, pelleted type materials prepared by compressing particles of catalyst powder have a bidisperse pore structure. The powder particles themselves are microporous, but the spaces between the powder particles are macropores. Usually micropores are defined as pores of radius less than about 10 nm and macropores are larger than this. In most cases the pore size distribution of micropores is measured by low temperature nitrogen adsorption [30] and that of macropores by mercury porosimetry [31]. In both methods, the pores are assumed to be interconnected cylindrical capillaries.

In 1962 Wakao and Smith [25] presented a model, which is known as the random pore model, for the estimation of effective diffusivity of a bidisperse pore structure. Figure 3.10 is a diagrammatic representation of the system used to describe the model. The dotted squares represent powder particles having micropores and the spaces between the squares represent the macropores. Writing the volume fractions of the macropores, micropores and solid as ϵ_a, ϵ_i and ϵ_s, respectively, gives

$$\epsilon_a + \epsilon_i + \epsilon_s = 1. \tag{3.72}$$

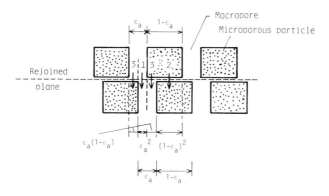

FIGURE 3.10   Random pore model of Wakao and Smith [25] for a bidisperse porous solid.

Suppose the sample is cut at a plane and the two surfaces are then rejoined. If the void area fraction on each of the two surfaces is ϵ, the void area fraction on the plane rejoined at random will be the possibility of two successive events, or ϵ^2. Therefore, the diffusion through a unit area of the rejoined plane can be divided into three additive parts, parallel mechanisms:

Mechanism 1   Diffusion through the macropores with an area of ϵ_a^2 and average pore radius $\bar{a}_a$.

Mechanism 2   Diffusion through the microporous particles having an area of $(1 - \epsilon_a)^2$ and an average pore radius $\bar{a}_i$.

Mechanism 3   Diffusion through the macropores and micropores in series. The area for this contribution is $2\epsilon_a(1 - \epsilon_a)$.

The diffusion rate per unit cross-sectional area of the porous media is then

---

| Mechanism 1 | Mechanism 2 | Mechanism 3 |

$$J_e = -\epsilon_a^2 D_a \frac{dc}{dx} - (1-\epsilon_a)^2 D_i \frac{dc}{dx} - 2\epsilon_a(1-\epsilon_a)\left(\frac{2}{\dfrac{1}{D_a}+\dfrac{1}{D_i}}\right)\frac{dc}{dx}.$$

$$(3.73)$$

The diffusivities, D_a and D_i, for the macropores and micropores, are given as an example for species 1 by

$$\left. \begin{array}{l} D_{1a} = \dfrac{1}{\dfrac{1-\alpha y_1}{D_{12}} + \dfrac{1}{\bar{D}_{K1,a}}} \\[3em] D_{1i} = \dfrac{1}{\dfrac{1-\alpha y_1}{D_{12}} + \dfrac{1}{\bar{D}_{K1,i}}} \end{array} \right\} \qquad (3.73a)$$

where $\bar{D}_{K1,a}$ and $\bar{D}_{K1,i}$ are the Knudsen diffusivities for species 1 in macropores and micropores, respectively: $\bar{D}_{K1,a} = 2\bar{v}_1\bar{a}_a/3$ and $\bar{D}_{K1,i} = 2\bar{v}_1\bar{a}_i/3$.

By evaluating $\bar{a}_a$ and $\bar{a}_i$ from the pore size distribution, Wakao and Smith showed that the diffusion fluxes, experimentally measured using alumina pellets in a He–N$_2$ system, were explained well by Eq. (3.73). The random pore model of Wakao and Smith was later extended by Cunningham and Geankoplis [32] to a more complicated tridisperse pore structure.

Foster and Butt [33] proposed a model for computing effective diffusivity from pore size distribution. The model considers the void volume in porous media to be composed of two major arrays of conical ducts, shown in Figure 3.11, which are made up of straight cylindrical capillary segments. One of the ducts is narrowest at the center of the solid and the other widest at the center. These two ducts, centrally convergent and

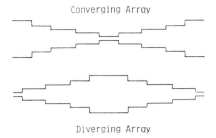

Converging Array

Diverging Array

FIGURE 3.11   Convergent–divergent pore array model of Foster and Butt [33].

centrally divergent, are the inverse of each other. The shape of these ducts is determined from the pore size distribution.

Diffusion flux through the ducts is estimated by trial and error calculation. By using an assumed value for the flux, the change in concentration across each capillary segment may be estimated from Eq. (3.50a) or its integration. For the i-th segment of length L_i, it is shown that

$$(y_1)_{i+1} = \left(1 + \frac{D_{12}}{D_{K1}}\right)\frac{1}{\alpha} + \left[(y_1)_i - \frac{1 + \dfrac{D_{12}}{D_{K1}}}{\alpha}\right]\exp\left(\frac{\alpha J_1 R_g T L_i}{D_{12} P}\right).$$

$$(3.74)$$

Starting with the concentration at one end of the solid sample, the concentration change is computed for each capillary segment. Mixing between the two ducts is assumed to take place at various points. The computation is carried out along the total length of the two ducts. The correct flux is then determined by comparing calculated and given concentrations at the end.

Johnson and Stewart [34] also presented a method for predicting the rate of diffusion through a porous solid. The rate of diffusion in each pore is calculated based on the dusty gas model of Evans et al. [17] and the total diffusion rate is then evaluated by integration over the entire range of pore size distribution. However, they stated that, because of possible anisotropy and other related effects, a diffusion or Knudsen permeability measurement was needed for accurate predictions.

3.3.2.3   *Effective diffusivity and surface diffusion*

For adsorbent particles, the contribution of pore volume diffusion may be estimated from pore size distribution. However, the effective diffusivity for an adsorption system consists of not only pore volume diffusion, but also pore surface migration. For a strong adsorbent, surface diffusion makes a larger contribution to the total transport.

A considerable amount of work has been reported on the measurement of surface diffusion in porous solids. In the earlier work of Babbit [35], and Gilliland *et al.* [36], a two-dimensional spreading pressure working on the adsorbed layer was regarded as the driving force for surface diffusion. However, in recent work, a model based on the random hopping of adsorbed molecules between adjacent sites has been employed more frequently, for example, in the work of Higashi *et al.* [37], Smith and Metzner [38], Weaver and Metzner [39], Gilliland *et al.* [40] and Ponzi *et al.* [41]. It was also pointed out by Thakur *et al.* [42] that the effect of collisions between gas molecules and mobile molecules in the adsorbed phase should be taken into account in evaluating surface diffusion flux.

However, it seems that at present a satisfactory prediction of surface diffusion for every gas–solid system is still far too difficult. Surface diffusion should be measured in a constant pressure Wicke–Kallenbach apparatus with both non-adsorbing (molecular weight M, diffusion flux N) and adsorbing (molecular weight M_a, diffusion flux N_a) gases. The pore volume diffusion flux measured for the non-adsorbing gas is then corrected for the molecular weight of the adsorbing species. Subtraction of this from the total diffusion flux measured for the adsorbing species gives the surface diffusion contribution of the adsorbing species:

$$\text{Surface diffusion} = N_a - N \left(\frac{M}{M_a} \right)^{1/2}.$$

If the permeability measurements are made under Knudsen diffusion conditions with an inert gas and an adsorbing gas separately, the surface diffusion of the adsorbing species may also be evaluated by the correction for the molecular weight ratio. The permeability, or forced-flow measurement, should be made completely in Knudsen diffusion region. Therefore, this method cannot be applied to porous solids with macropores. However, this restriction is not imposed on a constant pressure countercurrent diffusion measurement. Effective diffusivities comprising of pore volume and surface diffusion may be determined directly from a chromatography measurement as well. As a matter of fact, effective diffusivities of adsorbents, particularly of strong adsorbents, are easily determined from adsorption chromatography measurements at various flow rates.

3.3.2.4   *Effective diffusivity in multicomponent systems*

Rothfeld [19] has shown that Eq. (3.50a) can be extended to a multi-component system as:

$$J_m = -D_m \frac{\mathrm{d}c_m}{\mathrm{d}x} \tag{3.75}$$

where

$$D_m = \cfrac{1}{\cfrac{1}{D_{Km}} + \sum_n \cfrac{y_n - \left(\cfrac{J_n}{J_m}\right) y_m}{D_{mn}}} \tag{3.75a}$$

in which D_{mn} is the binary molecular diffusion coefficient for species m and n.

If a chemical reaction with the stoichiometry

$$a_1 A_1 + a_2 A_2 + \ldots a_m A_m + \ldots = 0 \tag{3.76}$$

where

$$a_m > 0 \quad \text{for reactants}$$

$$a_m < 0 \quad \text{for products}$$

proceeds in a porous catalyst, the reaction rates, r_v, or production rates of the reacting species are related as follows:

$$\frac{-(r_v)_1}{a_1} = \ldots = \frac{-(r_v)_m}{a_m} = \ldots = R_v. \tag{3.77}$$

R_v is often referred to as the reaction rate based on stoichiometry. The diffusion fluxes are related to the stoichiometric coefficients of the reaction, e.g.

$$\frac{J_n}{J_m} = \frac{a_n}{a_m}. \tag{3.78}$$

The diffusion flux in a porous solid having a monodisperse pore structure, for example, is then

$$J_{em} = -D_{em}\frac{dc_m}{dx}$$

(3.79)

where

$$D_{em} = \delta \; \frac{1}{\dfrac{1}{\bar{D}_{Km}} + \sum_n \dfrac{y_n - \left(\dfrac{a_n}{a_m}\right) y_m}{D_{mn}}}.$$

(3.79a)

Also, it should be noted that the diffusivity of m-th component through the so-called external film on a catalyst pellet, in which the chemical reaction of Eq. (3.76) takes place, is given by Eq. (3.75a) in conjunction with Eq. (3.78) and $D_{Km} = \infty$. Thus

$$(D_m)_{ext\;film} = \frac{1}{\sum_n \dfrac{y_n - \left(\dfrac{a_n}{a_m}\right) y_m}{D_{mn}}}.$$

(3.80)

There is a difference in the flux ratio between the two systems under reactive and inert conditions: the ratio is governed by Eq. (3.46) in constant pressure countercurrent diffusion and by Eq. (3.78) in diffusion with a chemical reaction. A question arises, i.e. are the two effective diffusivities under reactive and inert conditions the same, even if y_1, for example in a binary gas system, is so small that the flux ratio effect is of no significance. The answer is that they will be the same provided that the mean pore radius under the reaction conditions is identical to that under inert conditions.

Balder and Petersen [43] measured the reactive effective diffusivities from experiments on the hydrogenolysis of cyclopropane on a platinum/ alumina catalyst and found them to be almost the same as the non-reactive diffusivities measured for the same gas system. A similar conclusion was also reached by Toei et al. [44], who studied hydrogenation of ethylene on a nickel/diatomaceous earth catalyst.

Ryan *et al.* [45] developed a theory for the calculation of components of an effective diffusivity tensor under reaction conditions in a spatially periodic porous media. They showed that the effective diffusivities should be independent of the rate of chemical reaction.

3.4   Jüttner Modulus for First-order Reversible Reactions

Suppose a reversible chemical reaction,

$$aA \rightleftharpoons bB \tag{3.81}$$

with first-order kinetics,

$$(r_v)_A = -k_x\left(c_A - \frac{c_B}{K_{eq}}\right) \tag{3.82}$$

where K_{eq} is the equilibrium constant, takes place in the presence of inert component, I, in a spherical solid catalyst under isothermal conditions.

Writing Eq. (3.11) for each species as follows:

$$\frac{D_{eA}}{r^2}\frac{d}{dr}\left(r^2\frac{dc_A}{dr}\right) + (r_v)_A = 0 \tag{3.83}$$

$$\frac{D_{eB}}{r^2}\frac{d}{dr}\left(r^2\frac{dc_B}{dr}\right) + (r_v)_B = 0. \tag{3.84}$$

Also, from Eq. (3.77)

$$-\frac{(r_v)_A}{a} = \frac{(r_v)_B}{b} \tag{3.85}$$

therefore,

$$\frac{D_{eA}}{r^2}\frac{d}{dr}\left(r^2\frac{dc_A}{dr}\right) = -\frac{a}{b}\frac{D_{eB}}{r^2}\frac{d}{dr}\left(r^2\frac{dc_B}{dr}\right). \tag{3.85a}$$

If c_A, $c_B \ll c_I$, the effective diffusivities, D_{eA} and D_{eB}, are approximately independent of c_A and c_B. With these assumptions, Eq. (3.85a) is integrated to give

$$D_{eA}(c_A - C_{As}) = -\frac{a}{b}D_{eB}(c_B - C_{Bs}) \qquad (3.86)$$

where C_{As} and C_{Bs} are the concentrations of A and B, respectively, at the pellet surface.

The reaction rate is then rewritten as:

$$(r_v)_A = -k_x \left(1 + \frac{bD_{eA}}{aD_{eB}K_{eq}}\right)(c_A - c_{Ae}) \qquad (3.87)$$

where

$$c_{Ae} = \frac{C_{Bs} + \left(\dfrac{bD_{eA}}{aD_{eB}}\right)C_{As}}{K_{eq} + \dfrac{bD_{eA}}{aD_{eB}}}. \qquad (3.87a)$$

The Jüttner modulus is, therefore, modified as:

$$\phi_R = R\left[k_x\left(\frac{1}{D_{eA}} + \frac{b}{aD_{eB}K_{eq}}\right)\right]^{1/2}. \qquad (3.88)$$

The catalyst effectiveness factor for a first-order reversible reaction is evaluated from Eq. (3.19) with the modified modulus.

In the case of $a = b$ and $D_{eA} \simeq D_{eB}$, the modulus reduces to

$$\phi_R = R\left[\frac{k_x}{D_{eA}}\left(1 + \frac{1}{K_{eq}}\right)\right]^{1/2}. \qquad (3.89)$$

For catalyst pellets with a bidisperse macropore/micropore structure, Carberry [46] introduced a concept of macropore and micropore effectiveness factors, and developed a theory for an overall pellet effectiveness factor expressed in terms of the macropore and micropore effectiveness factors.

Example 3.1

Ethylene was hydrogenated at 130°C and 0.1 MPa in a 1.5 cm diameter differential reactor with fifty 0.26 cm spherical pellets of nickel/silica catalyst. The feed was a mixture of 10 cm^3 s^{-1} C$_2$H$_4$, 90 cm^3 s^{-1} H$_2$ and 100 cm^3 s^{-1} N$_2$ at 20°C and 0.1 MPa. The reaction product, ethane, in the exit stream was found to be 0.5%. The catalyst pellets are believed to have monodisperse pores of average diameter 50 nm. The intraparticle void fraction is 0.46. Estimate the catalyst effectiveness factor and intrinsic chemical reaction rate constant.

SOLUTION

Let us use the following notation:

	Molar flow rate	
Component	Reactor inlet	Reactor outlet
C$_2$H$_4$: A	F_{A1}	F_{A2}
H$_2$: B	F_{B1}	F_{B2}
C$_2$H$_6$: S	$F_{S1} = 0$	F_{S2}
N$_2$: I	F_{I1}	$F_{I2} = F_{I1}$
total	F_1	F_2

F_{v1} and F_{v2} are the volumetric flow rates at reactor inlet and outlet, respectively.

The reaction stoichiometry is

$$A + B = S. \qquad (3.90)$$

The changes in the number of moles are

$$F_{A1} - F_{A2} = F_{B1} - F_{B2} = F_{S2}$$

and then

$$F_2 = F_1 - F_{S2}.$$

(i) *Reactor inlet*

The inlet molar flow rates are

$$F_{A1} = \frac{(10)\,(273)}{(22{,}400)\,(293)} = 4.16 \times 10^{-4} \text{ mol s}^{-1}$$

$$F_{B1} = \frac{(90)\,(273)}{(22{,}400)\,(293)} = 37.4 \times 10^{-4} \text{ mol s}^{-1}$$

$$F_{I1} = \frac{(100)\,(273)}{(22{,}400)\,(293)} = 41.6 \times 10^{-4} \text{ mol s}^{-1}.$$

The total rate is

$$F_1 = (4.16 + 37.4 + 41.6) \times 10^{-4} = 83.2 \times 10^{-4} \text{ mol s}^{-1}.$$

The mole fractions are

$$y_{A1} = 0.05$$
$$y_{B1} = 0.45$$
$$y_{I1} = 0.50.$$

The volumetric flow rate at 130°C and 0.1 MPa is

$$F_{v1} = (10 + 90 + 100) \times 10^{-6} \times \frac{403}{293} = 2.75 \times 10^{-4} \text{ m}^3 \text{ s}^{-1}.$$

The ethylene concentration is

$$C_{A1} = \frac{4.16 \times 10^{-4}}{2.75 \times 10^{-4}} = 1.51 \text{ mol m}^{-3}.$$

(ii) *Reactor outlet*

$$\frac{F_{S2}}{F_2} = 0.005$$

$$F_2 = F_1 - F_{S2} = 83.2 \times 10^{-4} - F_{S2}$$

therefore,

$$F_{S2} = 0.414 \times 10^{-4} \text{ mol s}^{-1}$$
$$F_2 = 82.8 \times 10^{-4} \text{ mol s}^{-1}$$

and

$$F_{A2} = 3.75 \times 10^{-4} \text{ mol s}^{-1}$$
$$F_{B2} = 37.0 \times 10^{-4} \text{ mol s}^{-1}.$$

The mole fractions are

$$y_{A2} = \frac{F_{A2}}{F_2} = 0.045$$

$$y_{B2} = \frac{F_{B2}}{F_2} = 0.447$$

$$y_{S2} = 0.005$$

$$y_{I2} = \frac{F_{I1}}{F_2} = 0.502.$$

For the differential reactor it is clear that

$$\frac{F_2}{F_1} = \frac{F_{v2}}{F_{v1}}$$

therefore,

$$F_{v2} = F_{v1} \left(1 - \frac{F_{S2}}{F_1} \right)$$

$$= 2.75 \times 10^{-4} \left(1 - \frac{0.414 \times 10^{-4}}{83.2 \times 10^{-4}} \right) = 2.74 \times 10^{-4} \text{ m}^3 \text{ s}^{-1}.$$

The concentration of reactant A is

$$C_{A2} = \frac{F_{A2}}{F_{v2}} = \frac{3.75 \times 10^{-4}}{2.74 \times 10^{-4}} = 1.37 \text{ mol m}^{-3}.$$

(iii)   *Calculation of* $(R_p)_A$
The reaction rate throughout a single catalyst pellet is

$$(R_p)_A = \frac{F_{A2} - F_{A1}}{\text{number of pellets}}$$

$$= -\frac{0.41 \times 10^{-4}}{50} = -8.2 \times 10^{-7} \text{ mol s}^{-1}.$$

The average concentration of reactant A in the differential reactor is

$$(C_A)_{av} = \frac{C_{A1} + C_{A2}}{2}$$

$$= \frac{1.51 + 1.37}{2} = 1.44 \text{ mol m}^{-3}.$$

The volume of a single pellet is

$$\frac{4\pi R^3}{3} = \frac{4\pi (1.3 \times 10^{-3})^3}{3} = 9.2 \times 10^{-9} \text{ m}^3$$

therefore, from Eq. (3.22), we find

$$\frac{1}{k_x E_f} + \frac{R}{3k_f} = \frac{\left(\dfrac{4\pi R^3}{3}\right)(C_A)_{av}}{-(R_p)_A}$$

$$= \frac{(9.2 \times 10^{-9})(1.44)}{(8.2 \times 10^{-7})} = 0.016 \text{ s}.$$

(iv) *Estimation of k_f*

From Eq. (3.80) the diffusivity of ethylene in the external film of catalyst pellet for the reaction of Eq. (3.90) is

$$(D_A)_{\text{ext film}} = \cfrac{1}{\cfrac{y_B - y_A}{D_{AB}} + \cfrac{y_S + y_A}{D_{AS}} + \cfrac{y_I}{D_{AI}}} \cdot \tag{3.91}$$

The binary gas diffusivities at 130°C and 0.1 MPa are estimated to be

$$D_{AB} = 0.916 \times 10^{-4} \text{ m}^2 \text{ s}^{-1}$$
$$D_{AS} = 0.216 \times 10^{-4} \text{ m}^2 \text{ s}^{-1}$$
$$D_{AI} = 0.284 \times 10^{-4} \text{ m}^2 \text{ s}^{-1}.$$

The composition varies to some extent across the external film of a catalyst pellet. But, if we assume that the y values in Eq. (3.91) can be equated with the average mole fractions of the bulk gas in the reactor, then, the average mole fraction of gas A is

$$y_A = \frac{y_{A1} + y_{A2}}{2} = 0.048.$$

Similarly

$$y_B = 0.449$$
$$y_S = 0.0025$$

and

$$y_I = 0.501$$

therefore,

$$(D_A)_{\text{ext film}} = \cfrac{1}{\cfrac{0.449 - 0.048}{0.916 \times 10^{-4}} + \cfrac{0.0025 + 0.048}{0.216 \times 10^{-4}} + \cfrac{0.501}{0.284 \times 10^{-4}}}$$

$$= \frac{1}{(0.438 + 0.234 + 1.764) \times 10^4} = \frac{1}{2.436 \times 10^4}$$

$$= 4.11 \times 10^{-5} \text{ m}^2 \text{ s}^{-1}.$$

For calculating viscosity and density of the bulk gas in the reactor, the gas is assumed to be a mixture of H_2 and N_2 in the volume ratio 90:100. The viscosities of H_2 and N_2 at 130°C are $\mu_{H_2} = 1.10 \times 10^{-5}$ Pa s and $\mu_{N_2} = 2.24 \times 10^{-5}$ Pa s. From the chart of Bromley and Wilke [47], the viscosity of the gas mixture is then estimated to be $\mu = 2.15 \times 10^{-5}$ Pa s. The densities of H_2 and N_2 at 0°C and 0.1 MPa are $\rho_{H_2} = 0.0898$ kg m^{-3} and $\rho_{N_2} = 1.251$ kg m^{-3}, so that the density of the gas mixture at 130°C is

$$\rho_F = \left[\left(\frac{90}{190} \right) (0.0898) + \left(\frac{100}{190} \right) (1.251) \right] \frac{273}{403}$$

$$= 0.475 \text{ kg m}^{-3}.$$

The average volumetric flow rate is: $(F_{v1} + F_{v2})/2 = 2.745 \times 10^{-4} \text{ m}^3 \text{ s}^{-1}$. The superficial gas velocity is then

$$u = \frac{2.745 \times 10^{-4}}{\pi (0.75 \times 10^{-2})^2} = 1.55 \text{ m s}^{-1}$$

therefore,

$$Re = \frac{2Ru\rho_F}{\mu} = \frac{(2.6 \times 10^{-3})(1.55)(0.475)}{(2.15 \times 10^{-5})} = 89.0$$

and

$$Sc = \frac{\mu}{\rho_F (D_A)_{\text{ext film}}} = \frac{(2.15 \times 10^{-5})}{(0.475)(4.11 \times 10^{-5})} = 1.10.$$

From Eq. (4.11)

$$Sh = \frac{2k_f R}{(D_A)_{\text{ext film}}} = 2 + 1.1(1.10)^{1/3}(89.0)^{0.6} = 18.8$$

therefore,

$$k_f = \frac{(18.8)(4.11 \times 10^{-5})}{(2.6 \times 10^{-3})} = 0.297 \text{ m s}^{-1}.$$

(v) *Estimation of E_f and k_x*

$$\frac{1}{k_x E_f} = 0.016 - \frac{R}{3k_f} = 0.016 - \frac{(1.3 \times 10^{-3})}{(3)(0.297)}$$

$$= 0.016 - 0.00146 = 0.01454 \text{ s}$$

therefore,

$$k_x E_f = 68.8 \text{ s}^{-1}.$$

From Eq. (3.79a), the effective diffusivity of ethylene in the catalyst pellet is

$$D_{eA} = \delta \; \frac{1}{\dfrac{1}{\bar{D}_{KA}} + \dfrac{y_B - y_A}{D_{AB}} + \dfrac{y_S + y_A}{D_{AS}} + \dfrac{y_I}{D_{AI}}}. \qquad (3.92)$$

The Knudsen diffusivity of ethylene at 130°C in pores of radius $\bar{a} = 25$ nm is

$$\bar{D}_{KA} = 3.068\bar{a}\left(\frac{T}{M_A}\right)^{1/2}$$

$$= (3.068)(25 \times 10^{-9})\left(\frac{403}{0.028}\right)^{1/2} = 9.20 \times 10^{-6} \text{ m}^2 \text{ s}^{-1}.$$

Average mole fractions in the catalyst pellet should be used for the y values in Eq. (3.92). However, if in place of y we simply take the average mole fractions of the bulk gas in the reactor calculated in (iv), and if we assume $\delta = \epsilon_p^2$,

$$D_{eA} = (0.46)^2 \; \frac{1}{\dfrac{1}{9.20 \times 10^{-6}} + 2.436 \times 10^4} = 1.59 \times 10^{-6} \text{ m}^2 \text{ s}^{-1}$$

therefore,

$$(k_x E_f) \frac{R^2}{D_{eA}} = (68.8) \frac{(1.3 \times 10^{-3})^2}{1.59 \times 10^{-6}} = 73.1.$$

The left hand side is $E_f \phi^2$. If $\phi > 10$, $E_f = 3/\phi$, and $E_f \phi^2 = 3\phi$. Therefore,

$$\phi = \frac{73.1}{3} = 24.4.$$

In fact, this shows that the condition $\phi > 10$ is fulfilled. We can then see that the catalyst effectiveness factor, E_f, is $3/24.4 = 0.123$, and the intrinsic chemical reaction rate constant, k_x, is 560 s^{-1}.

REFERENCES

[1] A. Wheeler, *Advances in Catalysis*, Vol. 3, Academic Press, New York (1951).
[2] A. Wheeler, in *Catalysis*, edited by P. H. Emmett, Vol. 2, Reinhold, New York (1955).
[3] F. A. L. Dullien, *Porous Media: Fluid Transport and Pore Structure*, Academic Press, New York (1979).
[4] R. Jackson, *Transport in Porous Catalysis*, Elsevier, New York (1977).
[5] E. E. Petersen, *Chemical Reaction Analysis*, Prentice-Hall, New Jersey (1965).
[6] C. N. Satterfield, *Mass Transfer in Heterogeneous Catalysis*, MIT Press, Massachusetts (1970).
[7] J. M. Smith, *Chemical Engineering Kinetics*, 2nd edn., McGraw-Hill, New York (1970).
[8] G. R. Youngquist, *Ind. Eng. Chem.* **62** (No. 8), 52 (1970).
[9] F. Jüttner, *Z. Phys. Chemie* **65**, 595 (1909).
[10] E. W. Thiele, *Ind. Eng. Chem.* **31**, 916 (1939).
[11] G. Damköhler, *Der Chemieingenieur*, Vol. 3, Akadem. Verlag., Leipzig, p. 430 (1937).
[12] J. B. Zeldowitsch, *Acta Physicochim. URSS* **10**, 583 (1939).
[13] R. Aris, *Chem. Eng. Sci.* **6**, 262 (1957).
[14] S. Kasaoka and Y. Sakata, *Kagaku Kogaku* **31**, 164 (1967).
[15] T. Suzuki and T. Uchida, *J. Chem. Eng. Japan* **12**, 425 (1979).
[16] M. Knudsen, *Ann. Phys.* **28**, 75 (1909).
[17] R. B. Evans, G. M. Watson and E. A. Mason, *J. Chem. Phys.* **35**, 2076 (1961).
[18] D. S. Scott and F. A. L. Dullien, *AIChE J.* **8**, 113 (1962).
[19] L. B. Rothfeld, *AIChE J.* **9**, 19 (1963).
[20] C. H. Bosanquet, *British TA Rept. BR-507*, September 27 (1944), quoted in [21].
[21] W. G. Pollard and R. D. Present, *Phys. Rev.* **73**, 762 (1948).
[22] E. Wicke and R. Kallenbach, *Kolloid Z.* **97**, 135 (1941).
[23] P. B. Weisz, *Z. Phys. Chem.* **11**, 1 (1957).

[24] S. Kaguei, K. Matsumoto and N. Wakao, *Kagaku Kogaku Ronbunshu* **6**, 206 (1980).

[25] N. Wakao and J. M. Smith, *Chem. Eng. Sci.* **17**, 825 (1962).

[26] J. Hoogschagen, *Ind. Eng. Chem.* **47**, 906 (1955).

[27] J. P. Henry, B. Channakesavan and J. M. Smith, *AIChE J.* **7**, 10 (1961).

[28] R. B. Evans, J. Truitt and G. M. Watson, *J. Chem. Eng. Data* **6**, 522 (1961).

[29] D. S. Scott and K. E. Cox, *Can. J. Chem. Eng.* **3**, 201 (1960).

[30] R. P. Barrett, L. G. Joyner and P. P. Halenda, *J. Amer. Chem. Soc.* **73**, 373 (1951).

[31] H. L. Ritter and R. C. Drake, *Ind. Eng. Chem., Anal. Ed.* **17**, 787 (1945).

[32] R. S. Cunningham and C. J. Geankoplis, *Ind. Eng. Chem. Fund.* **7**, 535 (1968).

[33] R. N. Foster and J. B. Butt, *AIChE J.* **12**, 180 (1966).

[34] M. F. L. Johnson and W. E. Stewart, *J. Catal.* **4**, 248 (1965).

[35] J. D. Babbit, *Can. J. Res.* **28A**, 449 (1950).

[36] E. R. Gilliland, R. F. Baddour and J. L. Russel, *AIChE J.* **4**, 90 (1958).

[37] K. Higashi, H. Ito and J. Oishi, *J. Japan Atom. Energy Soc.* **5**, 846 (1963).

[38] R. K. Smith and A. B. Metzner, *J. Phys. Chem.* **68**, 2741 (1964).

[39] J. A. Weaver and A. B. Metzner, *AIChE J.* **12**, 655 (1966).

[40] E. R. Gilliland, R. F. Baddour, G. P. Perkinson and K. L. Sladek, *Ind. Eng. Chem. Fund.* **13**, 95 (1974).

[41] M. Ponzi, J. Papa, J. B. P. Rivarola and G. Zgrablich, *AIChE J.* **23**, 347 (1977).

[42] S. C. Thakur, C. F. Brown and G. L. Haller, *AIChE J.* **26**, 355 (1980).

[43] J. R. Balder and E. E. Petersen, *J. Catal.* **11**, 195 (1968).

[44] R. Toei, M. Okazaki, K. Nakanishi, Y. Kondo, M. Hayashi and Y. Shiozaki, *J. Chem. Eng. Japan* **6**, 50 (1973).

[45] D. Ryan, R. G. Carbonell and S. Whitaker, *Chem. Eng. Sci.* **35**, 10 (1980).

[46] J. J. Carberry, *AIChE J.* **8**, 557 (1962).

[47] L. A. Bromley and C. R. Wilke, *Ind. Eng. Chem.* **43**, 1641 (1951).

4 Particle-to-Fluid Mass Transfer Coefficients

ONE OF the important parameters needed in the design of packed bed systems is the particle-to-fluid mass transfer coefficient. In the past four decades, a substantial amount of work has been devoted to the study of this parameter.

Particle-to-fluid mass transfer studies were first carried out by Gamson *et al.* [1] and Hurt [2], both in 1943. They obtained mass transfer coefficients from measurements of the rates of evaporation of water from wet porous particles. Hurt [2] also reported mass transfer coefficients derived from the measurement of rates of naphthalene sublimation. Since their pioneering work, a large number of experimental studies have been carried out on mass transfer coefficients in packed bed systems.

Theoretical work has also been in progress. Pfeffer [3], and Pfeffer and Happel [4] applied a free surface cell model to the creeping flow region. Le Clair and Hamielec [5–7] proposed a zero vorticity cell model, and El-Kaissy and Homsy [8] applied the free surface cell model, zero vorticity cell model and distorted cell model to a multiparticle system at low Reynolds numbers. Nishimura and Ishii [9] also applied the free surface cell model to the study of mass transfer at high Reynolds numbers. These models which are based on different assumptions, generally give different and inconsistent values of particle-to-fluid mass transfer coefficients. Therefore, theoretical prediction of transfer coefficients is far from satisfactory.

When mass transfer occurs between a flowing fluid in a packed bed and the particle surface on which the concentration of the transferring species is constant, the resistance to mass transfer is considered to reside on the fluid side. In such a system, the unsteady mass balance equation of the transferring species, according to the dispersed plug flow model, may

138

be expressed as:

$$\frac{\partial C}{\partial t} = D_{ax}\frac{\partial^2 C}{\partial x^2} - U\frac{\partial C}{\partial x} - \frac{a}{\epsilon_b}k_f(C - C_{ps}) \qquad (4.1)$$

where

a = particle surface area per unit volume of packed bed
C = concentration of transferring species in the bulk fluid
C_{ps} = concentration of transferring species at the particle surface
D_{ax} = axial fluid dispersion coefficient
k_f = particle-to-fluid mass transfer coefficient
U = interstitial fluid velocity
ϵ_b = bed void fraction.

In the experimental measurements of mass transfer coefficients, most investigators have chosen to ignore the dispersion effect. For instance, in the experiments conducted by Satterfield and Resnick [10] on the catalytic decomposition of hydrogen peroxide in a packed bed of metal spheres, they obtained, for such a fast reaction, the mass transfer coefficients under the assumption of ideal plug flow, or no fluid dispersion in the bed. However, as discussed in Section 2.2 (Figure 2.7), the stagnant term, E°/D_v, of the dispersion coefficient for a fast reaction ($\phi = \infty$) is as large as 20. Under such conditions the axial fluid dispersion coefficient is given by Eq. (4.2):

$$\frac{\epsilon_b D_{ax}}{D_v} = 20 + 0.5\,(Sc)(Re) \qquad \text{for } Re > 5. \qquad (4.2)$$

As can be seen from Eq. (4.1), the values of the mass transfer coefficients obtained under the assumption of ideal plug flow ($D_{ax} = 0$), therefore, will be significantly different from those obtained using the large dispersion coefficients given by Eq. (4.2).

Evaporation, sublimation and dissolution follow the same sequence of steps as a fast chemical reaction on a particle surface. Mass transfer takes place between the bulk fluid and the particle surface where the mass transferring species is at a constant concentration, and no intraparticle diffusion is involved in the overall mass transfer. The bed in which this type of mass transfer proceeds should therefore have a dispersion coefficient as given by Eq. (4.2). Wakao and Funazkri [11] corrected the literature data for the axial fluid dispersion coefficient of Eq. (4.2) and obtained an empirical correlation for particle-to-fluid mass transfer coefficients.

A critical review of the published mass transfer coefficient data and their correction for axial dispersion effect are made in the following sections.

4.1   Review of the Published Gas Phase Data

The numerous packed bed mass transfer coefficients, reported in the literature, were obtained using various experimental methods under different conditions. For the purposes of data correlation, the following criteria have been adopted in the selection of the data:

a)   The particles in the bed are all active. Distended and diluted bed data are not considered.

b)   The number of particle layers in a mass transferring bed are greater than two.

Table 4.1 (for the gas phase) and Table 4.2 (for the liquid phase) list selected experimental work together with methods and operating conditions.

4.1.1   Evaporation of Water into an Air Stream: Steady-state Measurements

Since Gamson *et al.* [1] and Hurt [2] reported their results on the evapora-tion of water into air in 1943, the same system has been repeatedly studied by many investigators. Mass transfer coefficients were determined from the rate measurements during constant rate evaporation.

In the work of Gamson *et al.* [1] ($Re = 100$ to 4000), and Wilke and Hougen [12] ($Re = 45$ to 250), the particle surfaces were assumed to be at wet-bulb temperatures. Hurt [2] was the first to measure particle surface temperatures (for the two runs at $Re = 150$ and 370). Galloway *et al.* [13] ($Re = 150$ to 1200) found that the differences between the measured temperatures and the wet-bulb temperatures were less than $0.3°C$. Bradshaw and Myers [14] ($Re = 200$ to 4000) observed, in some of their experi-mental runs, that the surface temperatures were at wet-bulb values. However, De Acetis and Thodos [15] ($Re = 60$ to 2100) pointed out that considerable temperature differences existed between the measured surface temperatures and the wet-bulb values when flow rates were low. Since the experimental findings of De Acetis and Thodos, subsequent studies carried out by Thodos and coworkers [16–21] on the determination of transfer coefficients were all based on the measured surface temperatures.

In the work of Hougen *et al.* [1, 12], in which they assumed wet-bulb surface temperatures, there is some reservation about the reliability of their results, particularly those obtained at lower Reynolds numbers, $Re = 45$ to 150. Nevertheless, their data are not substantially different from those of Thodos *et al.*, which were determined based on experimentally measured surface temperatures. Therefore, all the data obtained from studies of water evaporation are included in the correction and correlation section later in this chapter, provided that the information on bed height and void fraction, required for correction for axial fluid dispersion, is given.

4.1.2 Evaporation of Organic Solvents: Steady-state Measurements

The determination of mass transfer coefficients from the rates of evaporation of organic solvents from particle surfaces into a stream of inert gases has been the subject of extensive investigation by Thodos and coworkers. The systems employed by Hobson and Thodos [16] are *n*-butanol, toluene, *n*-octane and *n*-dodecane in air, nitrogen, carbon dioxide and hydrogen; those of Petrovic and Thodos [19] are *n*-octane, *n*-decane, *n*-dodecane and *n*-tetradecane in air; and that of Wilkins and Thodos·[22] is *n*-decane in air. In these studies, the mass transfer coefficients were determined based on the measured temperatures of the particle surfaces under steady-state conditions.

4.1.3 Sublimation of Naphthalene: Steady-state Measurements

The rates of sublimation of naphthalene were measured by Hurt [2], Resnick and White [23], Chu *et al.* [24], and Bradshaw and Bennett [25] in their determinations of mass transfer coefficients. Compared to the liquid–gas system, the naphthalene–gas system has the advantage that the adiabatic temperature drop is small. But, the disadvantage of the system is that the vapor pressure of naphthalene has not been thoroughly investigated.

TABLE 4.1
Gas phase mass transfer experimental data[a].

Year	Investigator	Experimental method	Steady or unsteady state conditions	Material	Particle		Fluid	Sc	Re	Fluid dispersion considered
					Shape	Size (mm)				
1943	Gamson et al. [1]	Evaporation of water	Steady	Celite	Sphere	2.3, 3.0, 5.6, 8.4, 11.6	Air	0.61–0.62	100–4000	No
					Cylinder	4.1×4.8, 6.8×8.5, 9.8×11.7, 14.0×12.5, 18.8×16.9				
1943	Hurt [2]	Evaporation of water	Steady		Cylinder	9.5×9.5	Air	0.61	150 & 370	No
		Sublimation of naphthalene	Steady	Naphthalene	Cylinder	4.8×4.8, 9.5×9.5	Air H₂	2.5 4.0	7–670	
					Flake	2.0, 2.8, 4.1, 5.6				
1945	Wilke and Hougen [12]	Evaporation of water	Steady	Celite	Cylinder	3.1×3.1, 4.8×4.3, 6.6×7.2, 9.7×8.6, 13.4×12.8, 15.1×16.3, 18.2×16.9	Air	0.6	45–250	No
1949	Resnick and White [23][b]	Sublimation of naphthalene	Steady	Naphthalene	Granule (ground)	0.5, 0.8, 1.0, 1.1	Air CO₂ H₂	2.39 1.47 4.02	0.83–25	No
1951	Hobson and Thodos [16]	Evaporation of water and organic solvents	Steady	Porous packing	Sphere	9.4	Air, N₂, CO₂, H₂	0.61–5.1	8.6–330	No

Year	Reference	Process	Condition	Material	Shape	Size	Fluid	Sc	Re	
1953	Chu et al. [24]	Sublimation of naphthalene coated on particles	Steady	Glass beads, Lead shot, Celite, Rape seed	Sphere, Sphere, Cylinder	0.7, 0.7, 1.3, 2.0, 5.3, 5.5, 8.5, 13.7, 14.1, 2.0	Air	2.57	20–2000	No
1954	Satterfield and Resnick [10]	Decomposition of H_2O_2	Steady	Polished catalytic metal	Sphere	5.1	Vapor mixture of H_2O_2 and water	0.7–0.9	15–160	No
1957	Galloway et al. [13]	Evaporation of water	Steady	Celite	Sphere	17.1	Air	0.61	150–1200	No
1960	De Acetis and Thodos [15]	Evaporation of water	Steady	Celite	Sphere	15.9	Air	0.61	60–2100	No
1961	Bradshaw and Bennett [25]	Sublimation of naphthalene	Steady	Naphthalene	Sphere, Cylinder	9.5, 6.4, 9.5, 12.7	Air	2.57	440–9900	Yes
1963	McConnachie and Thodos [18]	Evaporation of water	Steady	Celite	Sphere	15.9	Air	0.61	100–2500^c	No
1963	Bradshaw and Myers [14]	Evaporation of water	Steady	Kaoline, AMT, Kaosorb, Celite	Sphere, Sphere, Cylinder, Cylinder	4.7, 8.8, 4.0×4.1, 4.2×4.2, 6.2×4.9	Air	0.6	400–6500^c	No
1963	Sen Gupta and Thodos [20]	Evaporation of water	Steady	Celite	Sphere	15.9	Air	0.61	800–2000	No
1964	Sen Gupta and Thodos [21]	Evaporation of water	Steady	Celite	Sphere	15.9	Air	0.61	2000–6000	No
1967	Malling and Thodos [17]	Evaporation of water	Steady	Celite	Sphere	15.7–15.9	Air	0.61	300–8500	Yes
1968	Petrovic and Thodos [19]	Evaporation of water and heavy hydrocarbons	Steady	Celite	Sphere	1.8, 2.2, 2.6, 3.1, 9.4	Air	0.6–5.45	3–230	Yes

TABLE 4.1 (Continued)

Year	Investigator	Experimental method	Steady or unsteady state conditions	Particle			Fluid	Sc	Re	Fluid dispersion considered
				Material	Shape	Size (mm)				
1969	Wilkins and Thodos [22]	Evaporation of n-decane	Steady	Celite	Sphere	2.6, 3.1	Air	3.72	150–180	Yes
1974	Wakao and Tanisho [27]	Pulse response, non-adsorption	Unsteady	Vanadium diatomaceous earth	Cylinder Granule (ground)	3.1×4.7 1.1	H_2	1.5	0.06–1.8	No
1976	Miyauchi et al. [28]	Pulse chromatography, chemical reaction	Unsteady	Porous packing	Sphere	0.7, 1.0, 1.1, 1.2, 1.4	C_3H_8, N_2, H_2, He	0.5–2.0	1–160	Yes
1976	Wakao et al. [29]	Pulse chromatography, adsorption	Unsteady	Activated carbon	Sphere	2.2	H_2	1.5	0.1–1.0	Yes
1977	Gangwal et al. [30]	Pulse chromatography, adsorption	Unsteady	Silica gel	Sphere	0.2	He	—	0.05–0.3	Yes

[a] Diluted beds, distended beds and data with a single particle layer are not included.

[b] Criticized by Bar-Ilan and Resnick [26].

[c] $Re = D_pG/\mu$ except for Refs. [14] and [18] where $Re = D_pG/[\mu(1-\epsilon_b)]$.

TABLE 4.2
Liquid phase mass transfer experimental data[a].

Year	Investigator	Experimental method	Steady or unsteady state conditions	Particle				Sc	Re	Fluid dispersion considered
				Material	Shape	Size (mm)	Fluid			
1949	Hobson and Thodos [31][b]	Extraction of *iso*-butanol and methyl ethyl ketone	Unsteady	Celite	Sphere	9.4, 16.1	Water	780–870	3–35	No
1949	McCune and Wilhelm [32]	Dissolution of 2-naphthol	Steady	2-naphthol	Sphere Flake	3.2, 4.8, 6.4 1.3, 2.1	Water	1190–1510	14–1770	No
1950	Gaffney and Drew [33]	Dissolution of organic acids	Steady	Succinic acid	Sphere Cylinder	6.4, 9.5, 12.4 6.3	Acetone *n*-butanol	160–180 10 100– 13 300	0.8–1480[c]	No
				Salicylic acid	Sphere Cylinder	9.6, 12.9 6.3	Benzene	340–430		
1951	Ishino et al. [34]	Dissolution of benzoic acid	Steady	Benzoic acid	Cylinder	3.9, 4.0, 4.1, 4.2, 4.3, 5.5, 6.2, 6.8	Water	1170–1610	1–180	No
1953	Evans and Gerald [35]	Dissolution of benzoic acid	Steady	Benzoic acid	Granule (ground)	0.6, 0.8, 1.4, 2.1	Water	990–1100	1–60	No
1953	Dryden et al. [36][d]	Dissolution of 2-naphthol and benzoic acid	Steady	2-naphthol Benzoic acid	Cylinder Cylinder	6.3 6.3	Water	810–1150	0.013–7.2[c]	No
1956	Dunn et al. [37]	Dissolution of lead	Steady	Lead	Sphere	2.0, 2.1, 2.2, 4.4	Mercury	120–140	32–1500[c]	No

TABLE 4.2 (Continued)

Year	Investigator	Experimental method	Steady or unsteady state conditions	Particle			Fluid	Sc	Re	Fluid dispersion considered
				Material	Shape	Size (mm)				
1956	Selke et al. [38]	Ion exchange	Unsteady	Amberlite IR-120	Sphere	0.4, 0.5, 0.6, 0.9	Copper sulfate solution	520 & 1130	2.7–120^c	No
1958	Wakao et al. [39]	Dissolution of 2-naphthol	Steady	2-naphthol	Cylinder	8.0, 8.1, 8.5	Water	1460–1760	0.4–3000	No
1963	Williamson et al. [40]	Dissolution of benzoic acid	Steady	Benzoic acid	Sphere	6.1, 6.3	Water	940–1140	0.04–53	No
1966	Wilson and Geankoplis [41]	Dissolution of benzoic acid	Steady	Benzoic acid	Sphere	6.4	Water 60%-propylene glycol	860–1100 52 300–70 600	0.0016–11	No
1969	Kasaoka and Nitta [42]	Dissolution of benzoic acid coated on particles	Steady	Steel	Sphere	2.8, 4.1, 6.4	Water, benzoic acid aqueous solution	350–2850	1–100	No
1975	Upadhyay and Tripathi [43]	Dissolution of benzoic acid	Steady	Benzoic acid	Cylinder	6.0, 7.7, 8.1, 8.6, 9.0, 11.2	Water	720–1350	2–2410^c	No

1975	Miyauchi et al. [44]	Pulse chromatography, chemical reaction	Unsteady	Sulfonic acid–ion exchange resin	Sphere	0.9, 1.5	Water	510–640	0.01–5	Yes
1976	Appel and Newman [45]	Cathodic reduction of ferricyanide to ferrocyanide	Steady	Stainless steel	Sphere	4.0	Mixture of aqueous solution of ferrocyanide, ferricyanide and potassium nitrate	1390–1450	0.008–0.17	No
1977	Kumar et al. [46]	Dissolution of benzoic acid	Steady	Benzoic acid	Cylinder	5.5 × 2.5, 8.8 × 3.4, 8.8 × 4.5, 9.6 × 2.8, 12.8 × 3.1, 12.8 × 3.8, 12.8 × 4.9	Water, 60% propylene glycol	770–42 400	0.01–600	No

[a] Diluted bed data are not included.

[b] Criticized by Gaffney and Drew [33], and Williamson et al. [40].

[c] $Re = D_p G/\mu$ except for Refs. [33] and [36–38] where $Re = D_p G/(\mu c_b)$, and Ref. [43] where $Re = D_p G/[\mu(1 - c_b)]$.

[d] Natural convection at $Re < 5c_b$.

TABLE 4.3
Vapor pressures of naphthalene reported in various sublimation studies.

Reference	Vapor pressure at 25°C (Pa)
Hurt [2]	15.3[a]
Resnick et al. [23]	11.1
Chu et al. [24][b]	11.7
Bradshaw et al. [25]	Not mentioned
Andrews [47]	9.64
Handbook of Chemistry and Physics [48]	11.7

[a] Value estimated by extrapolation, according to Resnick et al. [23].
[b] Data recommended in the International Critical Tables [49].

As compared in Table 4.3, different vapor pressures have been measured or assumed in different studies. The disagreement in the measured vapor pressures seems to have resulted in the disagreement in the obtained mass transfer coefficients. In fact, small errors in the vapor pressure value as well as in the measured outlet pressure are greatly magnified in the calculation of mass transfer coefficients, particularly when the difference between the two pressure values is small. This is very critical in the experimental determination of mass transfer coefficients, especially at low flow rates.

Table 4.3 indicates that quite high vapor pressures were assumed by Hurt [2]. This is probably the reason why he obtained relatively low transfer coefficients. Resnick and White [23] also obtained low transfer coefficients, which, according to Bar-Ilan and Resnick [26], are attributable to the improper experimental techniques employed in their measurements.

Thus, except for the transfer coefficients reported by Hurt and Resnick et al., all the other data mentioned above are included in the data correction and correlation.

4.1.4   Diffusion Controlled Catalytic Reaction on Particle Surfaces: Steady-state Measurements

Satterfield and Resnick [10] conducted experiments on the catalytic decomposition of hydrogen peroxide at a metal surface. The reaction is so fast that mass transfer between the particle surface and the bulk fluid is the rate controlling step. The mass transfer coefficients can be easily evaluated from measurements of the overall rates.

4.1.5 Pulse Gas Chromatography: Unsteady-state Measurements

Mass transfer coefficients in non-adsorption and adsorption systems were determined by Wakao and Tanisho [27] and Wakao *et al.* [29], respectively. The data were obtained from unsteady-state, pulse chromatography measurements. They obtained anomalously low transfer coefficients, but Wakao [50] has shown that the assumption of concentric intraparticle concentration inherent to the original Dispersion–Concentric model (in which solid phase mass diffusion in the axial direction is not considered) is responsible for the low coefficient values. Their original data, therefore, are not included in the data correlation.

Gangwal *et al.* [30] also made adsorption chromatography measurements and found the limiting particle-to-fluid mass transfer coefficient in terms of the Sherwood number being not less than unity.

Miyauchi *et al.* [28] made chromatography measurements of a chemical reaction between potassium hydroxide, presoaked on to particles, and carbon dioxide imposed as a pulse on a carrier gas flowing in a packed bed. They measured the overall mass transfer resistance, comprised of gas phase and solid phase diffusion resistances. They mentioned that the limiting Sherwood number was 12.5 at a bed void fraction of 0.5.

4.2 Review of the Published Liquid Phase Data

In some published data, the coefficients were determined based on measurements conducted at very low Reynolds numbers. The problem at low flow rates is that there is interference by natural convection. The convection effect becomes increasingly important as the Reynolds number decreases beyond a certain critical value. This critical Reynolds number, above which natural convection may be ignored is generally not clear. According to Dryden *et al.* [36], the critical Reynolds number for a bed packed with 6.3 mm particles is related to the bed void fraction, ϵ_b, by $Re \simeq 5\epsilon_b$.

In order to avoid the possible natural convection effect, the liquid phase data for $Re < 3$ are not considered in the data correlation.

4.2.1 Dissolution of a Solid into a Liquid Stream: Steady-state Measurements

Mass transfer coefficients determined from the rates of dissolution of solids into liquid streams are numerous. The data that will be considered for the correlation are obtained from the following work.

The rates of dissolution of spherical or cylindrical particles of benzoic acid into water flowing through the beds were measured by Ishino *et al.* [34], Dryden *et al.* [36], Williamson *et al.* [40], Wilson and Geankoplis [41], Kasaoka and Nitta [42], Upadhyay and Tripathi [43], Kumar *et al.* [46]. Evans and Gerald [35] used finely ground granular particles of benzoic acid in their measurements of rate of dissolution. Dissolution of benzoic acid into propylene glycol was studied by Wilson and Geankoplis [41], and Kumar *et al.* [46].

Dissolution of 2-naphthol into water was investigated by McCune and Wilhelm [32], Dryden *et al.* [36], and Wakao *et al.* [39]. Dissolution of succinic acid and salicylic acid into acetone, *n*-butanol and benzene was studied by Gaffney and Drew [33]. Dissolution of lead into mercury was investigated by Dunn *et al.* [37].

4.2.2 Electrochemical Reaction: Steady-state Measurements

Appel and Newman [45] applied a limiting current method to obtain the mass transfer coefficients at low flow rates: $Re \lesssim 0.17$.

4.2.3 Extraction of Liquids: Unsteady-state Measurements

Hobson and Thodos [31] conducted experiments on the extraction of *iso*-butanol and methyl ethyl ketone, presoaked on to porous Celite particles, into water flowing through the bed. They measured the variation with time of the effluent concentration and evaluated the initial extraction rate from the extrapolation of the curve, having a rapidly changing slope, to "zero time". However, the work has been criticized by Gaffney and Drew [33], and Williamson *et al.* [40] for the uncertainty involved in the evaluation of initial rates by the method of extrapolation. Therefore, the data obtained by Hobson and Thodos are not used in the data correlation.

4.2.4 Ion Exchange: Unsteady-state Measurements

Ion exchange in Amberlite particles was studied by Selke *et al.* [38]. Their mass transfer coefficients, determined in the Reynolds number range 1 to 40, are considerably larger than those obtained by many other investigators. The graphical method they applied for the determination of transfer coefficients does not seem to give accurate coefficient values. Their data will not be included in the correlation.

4.2.5 Pulse Liquid Chromatography: Unsteady-state Measurements

Miyauchi *et al.* [44] carried out chromatography measurements of the reaction between sodium hydroxide, imposed as a pulse on a stream of water, and sulfonic acid in ion exchange resin particles. They determined the overall mass transfer resistance, employing similar techniques to those used in the gas chromatography measurements [28]. They reported that the limiting Sherwood number was 16.7 at a bed void fraction of 0.4.

4.3   Re-evaluation of the Mass Transfer Data

As a result of the analysis in the preceding sections, the data which have satisfied the requirements for the correlation are as follows:

a)   Evaporation of water [1, 2, 12-21];

b)   Evaporation of organic solvents [16, 19, 22];

c)   Sublimation of naphthalene [24, 25];

d)   Diffusion controlled reaction on particle surfaces [10];

e)   Dissolution of solids [32-37, 39-43, 46].

The dissolution data are available for a very wide range of Reynolds number, 0.0016 to 3000. But, to avoid any possible natural convection effect, as mentioned already, the data [36] at Reynolds number less than about three are not included in the data correlation.

Incidentally, all the data selected are those obtained under steady-state conditions, with solid particles having a constant concentration of mass transferring species at the surface. The data obtained from unsteady-state measurements have not passed the criteria. In general, two rate parameters are involved in the analysis of steady-state measurements: the particle-to-fluid mass transfer coefficient and the fluid dispersion coefficient. In the analysis of unsteady-state mass transfer by pulse chromatography and ion exchange, additional parameters, such as intraparticle diffusivity and intra-particle void fraction, are involved. It is conceivable that this makes the determination of the transfer coefficients in unsteady-state processes more complicated.

Under steady-state conditions Eq. (4.1) can be rewritten as:

$$U \frac{dC}{dx} - D_{ax} \frac{d^2C}{dx^2} + \frac{a}{\epsilon_b} k_f(C - C_{ps}) = 0. \qquad (4.3)$$

The following two types of packed beds have been used, so far, for the mass transfer measurements:

a)   Empty column before the mass transferring packed bed $(0 < x < L)$

b)   Inert packed bed $(-l < x < 0$; concentration $C')$ before the mass transferring packed bed $(0 < x < L$; concentration $C)$.

In type (a), the Danckwerts boundary conditions are used

$$\left. \begin{aligned} U(C - C_{in}) &= D_{ax} \frac{dC}{dx} && \text{at } x = 0 \\[2mm] \frac{dC}{dx} &= 0 && \text{at } x = L. \end{aligned} \right\} \qquad (4.4)$$

In type (b), same-sized particles are usually packed in both beds, but the fluid dispersion coefficient, $(D_{ax})_{inert}$, in the inert bed may be different from D_{ax} in the mass transferring bed. The system is described by the following conditions:

$$\left. \begin{aligned} U(C' - C_{in}) &= (D_{ax})_{inert} \frac{dC'}{dx} && \text{at } x = -l \\[2mm] U \frac{dC'}{dx} - (D_{ax})_{inert} \frac{d^2C'}{dx^2} &= 0 && \text{for } -l < x < 0 \\[2mm] C' &= C && \text{at } x = 0 \\[2mm] (D_{ax})_{inert} \frac{dC'}{dx} &= D_{ax} \frac{dC}{dx} && \text{at } x = 0 \\[2mm] \frac{dC}{dx} &= 0 && \text{at } x = L. \end{aligned} \right\} \qquad (4.5)$$

Following the discussion of Wehner and Wilhelm [51], outlined in Section 2.1, it is easily shown that both types (a) and (b) give the same effluent concentration:

$$\frac{C_{ps} - C_{exit}}{C_{ps} - C_{in}} = \frac{4A \exp\left(\dfrac{LU}{2D_{ax}}\right)}{(1 + A)^2 \exp\left(A\dfrac{LU}{2D_{ax}}\right) - (1 - A)^2 \exp\left(-A\dfrac{LU}{2D_{ax}}\right)} \quad (4.6)$$

where

$$A = \left(1 + \frac{4k_f a D_{ax}}{\epsilon_b U^2}\right)^{1/2}. \quad (4.6a)$$

If $D_{ax} \to 0$, Eq. (4.6) reduces to

$$\frac{C_{ps} - C_{exit}}{C_{ps} - C_{in}} = \exp\left[-\frac{Sh^\dagger}{(Sc)(Re)} aL\right] \quad (4.7)$$

where $Sh^\dagger$ is a Sherwood number evaluated under the assumption of $D_{ax} = 0$.

Our main concern is about the mass transfer coefficients re-evaluated with D_{ax} given by either Eq. (2.26), in its general form, or by Eq. (4.2) for the mass transfer systems under consideration. The Sherwood number based on the re-evaluated mass transfer coefficient is denoted by Sh.

From the $Sh^\dagger$ data reported in the literature, it is feasible to calculate Sh values by equating Eqs. (4.6) and (4.7) when information on the bed height (or number of particle layers) and bed void fraction is given. Mass transfer coefficients have been obtained by some researchers [17, 19, 22, 25] by assuming that the axial Peclet number equals two. These data are also easily converted into Sh values.

The $Sh^\dagger$ values are re-evaluated for all the steady-state measurements listed in the preceding section except for those given by Gamson *et al.* [1], Hurt [2], and Bradshaw and Myers [14] who gave no detailed data about L and/or ϵ_b.

The Sh values recalculated from the data of Satterfield and Resnick [10], and Petrovic and Thodos [19] are shown together with their $Sh^\dagger$ data in Figure 4.1. A considerable difference can be seen between the two Sherwood values. The difference increases with a decrease in Reynolds number.

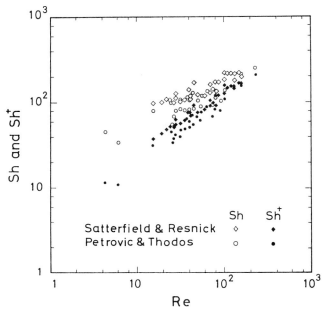

FIGURE 4.1 $Sh^\dagger$ of Satterfield and Resnick [10] and Petrovic and Thodos [19] and Sh values corrected for fluid dispersion.

Figure 4.2 illustrates the Sh values re-evaluated from the evaporation data with a Schmidt number, Sc, of 0.6. For mass (and heat) transfer of a single sphere, the following Ranz and Marshall [52] equation is popularly recognized:

$$Sh = 2 + 0.6Sc^{1/3}Re^{1/2}. \tag{4.8}$$

Figure 4.2 shows that the Sh values for packed beds are generally higher than those predicted from Eq. (4.8) for single spheres (dashed line), but the difference diminishes as the Reynolds number is lowered. It seems that with the decrease in Reynolds number, the Sherwood number for packed beds reduces and approaches the same limiting value of two for single spheres. Accordingly, the following empirical equation may be assumed:

$$Sh = 2 + \alpha Sc^{1/3}Re^{\beta}. \tag{4.9}$$

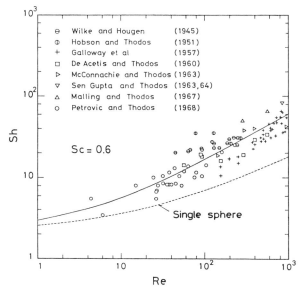

FIGURE 4.2   *Sh* versus *Re*, for water evaporation (solid line showing Eq. 4.11).

It should be noted that Petrovic and Thodos [19] have obtained an equation for the gas phase data corrected for an axial Peclet number of two:

$$\epsilon_b J_{Mass} = 0.357 Re^{-0.359} \qquad \text{for } 3 < Re < 900. \qquad (4.10)$$

At high Reynolds numbers, the data shown in Figure 4.2 are satisfactorily represented by the Petrovic–Thodos relationship as well. This is quite natural since, at high Reynolds numbers, the second term on the right-hand side of Eq. (4.2) is dominant: axial Peclet number being two. However, at lower Reynolds numbers, the re-evaluated data are higher than those according to Eq. (4.10). As mentioned already, the correction for larger axial dispersion coefficients gives higher mass transfer coefficients, particularly at low flow rates.

In a liquid phase system, *Sh* values are relatively large so that the liquid phase data are good to use for the determination of α and β values. Figure 4.3 is a plot of the liquid phase *Sh* data as $(Sh - 2)/Sc^{1/3}$ against *Re*. In the

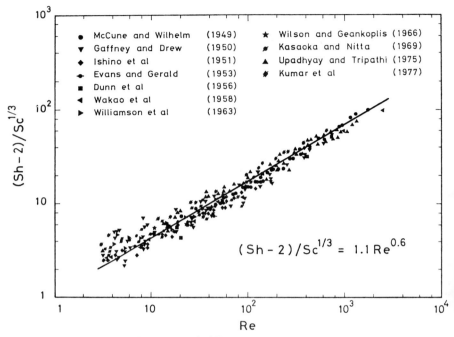

FIGURE 4.3 $(Sh - 2)/Sc^{1/3}$ versus Re, for liquid phase data.

wide range of Reynolds number from about 3 to 3000, the Sherwood data fit well a single straight line corresponding to the following correlation:

$$Sh = 2 + 1.1Sc^{1/3}Re^{0.6}. \qquad (4.11)$$

Equation (4.6) reduces to Eq. (4.7) when $A \to 1$, i.e. $D_{ax} \to 0$ or $k_f/(\epsilon_b U) = Sh/(Sc\,Re)$ is low. This indicates that liquid phase (high Sc) coefficients are affected little by the axial fluid dispersion unless Re is very low, whereas gas phase (low Sc) coefficients are considerably affected by the fluid dispersion, particularly at low Reynolds numbers.

In Figure 4.4, both liquid phase and gas phase Sh data are plotted against $(Sc^{1/3}Re^{0.6})^2$. The reason the square of $Sc^{1/3}Re^{0.6}$ is taken is in order to enlarge the plot in the x-axis direction. It is seen that the data are well correlated by Eq. (4.11), represented by the solid line.

In Figure 4.5, the range of the recalculated Sh values are compared with

GAS-PHASE		
⊖ Wilke and Hougen	(1945)	
⊕ Hobson and Thodos	(1951)	
◁ Chu *et al.*	(1953)	
◇ Satterfield and Resnick	(1954)	
+ Galloway *et al.*	(1957)	
▫ De Acetis and Thodos	(1960)	
⊠ Bradshaw and Bennett	(1961)	
▷ McConnachie and Thodos	(1963)	
▽ Sen Gupta and Thodos	(1963, 64)	
△ Malling and Thodos	(1967)	
○ Petrovic and Thodos	(1968)	
☆ Wilkins and Thodos	(1969)	

LIQUID-PHASE		
• McCune and Wilhelm	(1949)	
▾ Gaffney and Drew	(1950)	
✦ Ishino *et al.*	(1951)	
➔ Evans and Gerald	(1953)	
■ Dunn *et al.*	(1956)	
◄ Wakao *et al.*	(1958)	
► Williamson *et al.*	(1963)	
★ Wilson and Geankoplis	(1966)	
⚹ Kasaoka and Nitta	(1969)	
▲ Upadhyay and Tripathi	(1975)	
✶ Kumar *et al.*	(1977)	

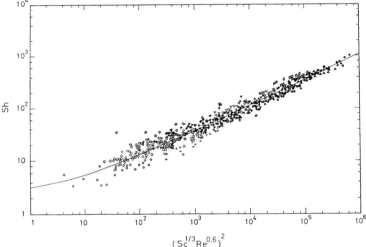

FIGURE 4.4   Correlation of Sherwood numbers, Sh, for gas and liquid phase data.

the corresponding $Sh^\dagger$ data. When $Sc^{1/3}Re^{0.6}$ is large, the Sh values are only slightly higher than the $Sh^\dagger$ data. As $Sc^{1/3}Re^{0.6}$ decreases, however, the difference between the two Sherwood values becomes more prominent and then significant at low $Sc^{1/3}Re^{0.6}$.

In many studies, the mass transfer coefficients determined, assuming $D_{ax} = 0$, have been correlated in terms of the J_{Mass} factor (see Dwivedi and Upadhyay [53], for instance). The fact that the $Sh^\dagger$ values decrease with decreasing Reynolds number appears as if a J_{Mass} factor correlation were valid even at very low flow rates. However, this is not exactly correct because recalculated Sh values are found to approach a limiting value.

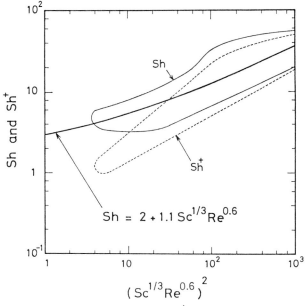

FIGURE 4.5   Comparison of Sh and $Sh^{\dagger}$, for gas and liquid phase data.

There is somewhat of an uncertainty about the limiting Sherwood value of two. The limiting value may be higher or lower than two. However, it is not likely that particle-to-fluid mass transfer is the rate controlling step at low flow rates. It seems that a Sherwood number greater than a certain value, say $Sh \gtrsim 1$, will explain mass transfer phenomena in packed beds at low flow rates. The important thing is that the Sherwood number does not keep decreasing with a decrease in Reynolds number. In fact, if the Sherwood number keeps decreasing, no mass transfer will occur between a particle and its surrounding fluid at very low or zero flow rates. This is certainly against the nature of mass transfer in packed beds.

REFERENCES

[1]   B. W. Gamson, G. Thodos and O. A. Hougen, *Trans. Amer. Inst. Chem. Eng.* **39**, 1 (1943).
[2]   D. M. Hurt, *Ind. Eng. Chem.* **35**, 522 (1943).
[3]   R. Pfeffer, *Ind. Eng. Chem. Fund.* **3**, 380 (1964).

[4]   R. Pfeffer and J. Happel, *AIChE J.* **10**, 605 (1964).
[5]   B. P. Le Clair and A. E. Hamielec, *Ind. Eng. Chem. Fund.* **7**, 542 (1968).
[6]   B. P. Le Clair and A. E. Hamielec, *Ind. Eng. Chem. Fund.* **9**, 608 (1970).
[7]   B. P. Le Clair and A. E. Hamielec, *Can. J. Chem. Eng.* **49**, 713 (1971).
[8]   M. M. El-Kaissy and G. M. Homsy, *Ind. Eng. Chem. Fund.* **12**, 82 (1973).
[9]   Y. Nishimura and T. Ishii, *Chem. Eng. Sci.* **35**, 1205 (1980).
[10] C. N. Satterfield and H. Resnick, *Chem. Eng. Prog.* **50**, 504 (1954).
[11] N. Wakao and T. Funazkri, *Chem. Eng. Sci.* **33**, 1375 (1978).
[12] C. R. Wilke and O. A. Hougen, *Trans. Amer. Inst. Chem. Eng.* **41**, 445 (1945).
[13] L. R. Galloway, W. Komarnicky and N. Epstein, *Can. J. Chem. Eng.* **35**, 139 (1957).
[14] R. D. Bradshaw and J. E. Myers, *AIChE J.* **9**, 590 (1963).
[15] J. De Acetis and G. Thodos, *Ind. Eng. Chem.* **52**, 1003 (1960).
[16] M. Hobson and G. Thodos, *Chem. Eng. Prog.* **47**, 370 (1951).
[17] G. F. Malling and G. Thodos, *Int. J. Heat Mass Trans.* **10**, 489 (1967).
[18] J. T. L. McConnachie and G. Thodos, *AIChE J.* **9**, 60 (1963).
[19] L. J. Petrovic and G. Thodos, *Ind. Eng. Chem. Fund.* **7**, 274 (1968).
[20] A. Sen Gupta and G. Thodos, *AIChE J.* **9**, 751 (1963).
[21] A. Sen Gupta and G. Thodos, *Ind. Eng. Chem. Fund.* **3**, 218 (1964).
[22] G. S. Wilkins and G. Thodos, *AIChE J.* **15**, 47 (1969).
[23] W. Resnick and R. R. White, *Chem. Eng. Prog.* **45**, 377 (1949).
[24] J. C. Chu, J. Kalil and W. A. Wetteroth, *Chem. Eng. Prog.* **49**, 141 (1953).
[25] R. D. Bradshaw and C. O. Bennett, *AIChE J.* **7**, 48 (1961).
[26] M. Bar-Ilan and W. Resnick, *Ind. Eng. Chem.* **49**, 313 (1957).
[27] N. Wakao and S. Tanisho, *Chem. Eng. Sci.* **29**, 1991 (1974).
[28] T. Miyauchi, H. Kataoka and T. Kikuchi, *Chem. Eng. Sci.* **31**, 9 (1976).
[29] N. Wakao, K. Tanaka and H. Nagai, *Chem. Eng. Sci.* **31**, 1109 (1976).
[30] S. K. Gangwal, R. R. Hudgins and P. L. Silveston, The Limiting Sherwood Number for Mass Transfer in Packed Beds, *Proceedings of the European Congress on Transfer Processes in Particle Systems*, Nürnberg, March (1977).
[31] M. Hobson and G. Thodos, *Chem. Eng. Prog.* **45**, 517 (1949).
[32] L. K. McCune and R. H. Wilhelm, *Ind. Eng. Chem.* **41**, 1124 (1949).
[33] B. J. Gaffney and T. B. Drew, *Ind. Eng. Chem.* **42**, 1120 (1950).
[34] T. Ishino, T. Otake and T. Okada, *Kagaku Kogaku* **15**, 255 (1951).
[35] G. C. Evans and C. F. Gerald, *Chem. Eng. Prog.* **49**, 135 (1953).
[36] C. E. Dryden, D. A. Strang and A. E. Withrow, *Chem. Eng. Prog.* **49**, 191 (1953).
[37] W. E. Dunn, C. F. Bonilla, C. Ferstenberg and B. Gross, *AIChE J.* **2**, 184 (1956).
[38] W. A. Selke, Y. Bard, A. D. Pasternak and S. K. Aditya, *AIChE J.* **2**, 468 (1956).
[39] N. Wakao, T. Oshima and S. Yagi, *Kagaku Kogaku* **22**, 780 (1958).
[40] J. E. Williamson, K. E. Bazaire and C. J. Geankoplis, *Ind. Eng. Chem. Fund.* **2**, 126 (1963).
[41] E. J. Wilson and C. J. Geankoplis, *Ind. Eng. Chem. Fund.* **5**, 9 (1966).
[42] S. Kasaoka and K. Nitta, *Kagaku Kogaku* **33**, 1231 (1969).
[43] S. N. Upadhyay and G. Tripathi, *J. Chem. Eng. Data* **20**, 20 (1975).
[44] T. Miyauchi, K. Matsumoto and T. Yoshida, *J. Chem. Eng. Japan* **8**, 228 (1975).
[45] P. W. Appel and J. Newman, *AIChE J.* **22**, 979 (1976).
[46] S. Kumar, S. N. Upadhyay and V. K. Mathur, *Ind. Eng. Chem. Process Des. Dev.* **16**, 1 (1977).

[47] M. R. Andrews, *J. Phys. Chem.* **30**, 1497 (1926).
[48] *Handbook of Chemistry and Physics*, edited by R. C. Weast, 53rd edn., Chemical Rubber Company, Ohio, D-164 (1972–73).
[49] *International Critical Tables*, Vol. 3, McGraw-Hill, New York, p. 208 (1928).
[50] N. Wakao, *Chem. Eng. Sci.* **31**, 1115 (1976).
[51] J. F. Wehner and R. H. Wilhelm, *Chem. Eng. Sci.* **6**, 89 (1956).
[52] W. E. Ranz and W. R. Marshall, *Chem. Eng. Prog.* **48**, 173 (1952).
[53] P. N. Dwivedi and S. N. Upadhyay, *Ind. Eng. Chem. Process Des. Dev.* **16**, 157 (1977).

5 Steady-State Heat Transfer

IN THIS chapter, a packed bed used as a heat exchanger will be considered. Fluid may be heated or cooled from the column wall while flowing through a packed bed. In such a packed bed operated under steady-state conditions, a difference in local temperature between the fluid and the particle may exist, but the overall solid and fluid temperature profiles are considered to be identical to each other, as sketched in Figure 5.1. In estimating the overall steady-state temperature profiles, the heterogeneous packed bed may be assumed to be a homogeneous single phase. The temperature profiles in the bed are then predicted in terms of effective thermal conductivities and wall heat transfer coefficients.

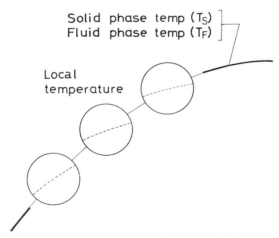

FIGURE 5.1   Steady-state temperature profiles in a packed bed (of heat exchanger type).

In the following sections, analytical solutions of the steady-state temperature profiles under different conditions will be shown. In addition, the various empirical formulae proposed for the estimation of effective thermal conductivities as well as wall heat transfer coefficients are discussed.

5.1   Steady-state Bed Temperature

For a cylindrical packed bed operated as a steady-state heat exchanger, the heat balance equation gives

$$GC_F \frac{\partial T}{\partial x} = k_{er} \frac{1}{r} \frac{\partial}{\partial r} \left(r \frac{\partial T}{\partial r} \right) + k_{eax} \frac{\partial^2 T}{\partial x^2} \tag{5.1}$$

where

C_F = specific heat of fluid at constant pressure
G = fluid mass velocity per unit area of bed cross-section
k_{eax} = effective axial thermal conductivity
k_{er} = effective radial thermal conductivity
T = bed temperature.

At intermediate and high flow rates, the axial second derivative, $\partial^2 T/\partial x^2$, is very small compared to the other terms. Equation (5.1) then reduces to

$$GC_F \frac{\partial T}{\partial x} = k_{er} \frac{1}{r} \frac{\partial}{\partial r} \left(r \frac{\partial T}{\partial r} \right). \tag{5.2}$$

Equation (5.2) was first solved by Hatta and Maeda [1] in 1948, and a few years later, by Coberly and Marshall [2], both based on the following boundary conditions:

$$\left. \begin{array}{ll} T = T_0 & \text{at } x = 0 \\[2ex] \dfrac{\partial T}{\partial r} = 0 & \text{at } r = 0 \\[2ex] k_{er} \dfrac{\partial T}{\partial r} = h_w(T_w - T) & \text{at } r = R_T \end{array} \right\} \tag{5.3}$$

where

h_w = wall heat transfer coefficient
R_T = column radius
T_0 = bed inlet temperature
T_w = wall temperature.

The solution to Eq. (5.2) with the conditions given by Eq. (5.3) is:

$$\frac{T_w - T}{T_w - T_0} = 2 \sum_{n=1}^{\infty} \frac{J_0(a_n r/R_T) \exp(-a_n^2 y)}{a_n[1 + (a_n/B)^2] J_1(a_n)} \tag{5.4}$$

where

$$B = \frac{h_w R_T}{k_{er}} \tag{5.4a}$$

$$y = \frac{k_{er} x}{G C_F R_T^2} \tag{5.4b}$$

and a_n is an n-th root of the following equation of Bessel functions (J_0 is a Bessel function of first kind and zeroth-order, and J_1 is that of first kind and first-order):

$$B J_0(a_n) = a_n J_1(a_n). \tag{5.4c}$$

5.1.1   Solution Deep in a Bed

When y, as defined by Eq. (5.4b), is greater than about 0.2, the series in Eq. (5.4) converges so rapidly such that only the first term of the series is significant. Therefore,

$$\frac{T_w - T}{T_w - T_0} = \frac{2 J_0(a_1 r/R_T) \exp(-a_1^2 y)}{a_1[1 + (a_1/B)^2] J_1(a_1)} \tag{5.5}$$

where

$$B J_0(a_1) = a_1 J_1(a_1). \tag{5.5a}$$

Equation (5.5) gives the temperature profile deep in the bed.
At $r = 0$ ($T = T_c$), Eq. (5.5) reduces to

$$\frac{T_w - T_c}{T_w - T_0} = \frac{2 \exp(-a_1^2 y)}{a_1[1 + (a_1/B)^2] J_1(a_1)} \tag{5.6}$$

which gives the temperature profile along the central axis of the bed under the conditions specified.

5.1.2 Mixed Mean Temperature of the Fluid

The mixed mean temperature, T_m, is the average radial temperature defined as:

$$T_m = \frac{2}{R_T^2} \int_0^{R_T} T(r) r \, dr. \tag{5.7}$$

Hence, Eq. (5.4), when expressed in terms of T_m, becomes

$$\frac{T_w - T_m}{T_w - T_0} = 4 \sum_{n=1}^{\infty} \frac{\exp(-a_n^2 y)}{a_n^2[1 + (a_n/B)^2]}. \tag{5.8}$$

For $y > 0.2$, the above equation converges to

$$\frac{T_w - T_m}{T_w - T_0} = \frac{4 \exp(-a_1^2 y)}{a_1^2[1 + (a_1/B)^2]}. \tag{5.9}$$

In terms of the radial mixed mean temperature the steady-state heat balance equation of the heat exchanger is

$$GC_F \frac{dT_m}{dx} = \frac{2U_0}{R_T}(T_w - T_m) \tag{5.10}$$

where U_0 is the overall heat transfer coefficient between the wall and the bed. The solution to Eq. (5.10) is

$$\frac{T_{\mathrm{w}} - T_{\mathrm{m}}}{T_{\mathrm{w}} - T_0} = \exp\left(-\frac{2U_0 x}{GC_{\mathrm{F}}R_{\mathrm{T}}}\right). \tag{5.11}$$

At sufficiently large values of x, comparison of Eqs. (5.9) and (5.11) gives

$$\frac{U_0 D_{\mathrm{T}}}{k_{\mathrm{F}}} = a_1^2 \frac{k_{\mathrm{er}}}{k_{\mathrm{F}}} \tag{5.12}$$

where D_{T} is the column diameter or $2R_{\mathrm{T}}$, and k_{F} is the thermal conductivity of the fluid flowing in the bed.

5.1.3   Solution when the Bed Inlet Temperature is a Function of Radial Distance

With a radial temperature profile at the bed inlet, i.e.

$$T = T_0(r) \qquad \text{at } x = 0$$

the solution to Eq. (5.2), when axial heat conduction is ignored, is

$$\frac{T_{\mathrm{w}} - T}{T_{\mathrm{w}} - T_0(0)} = 2 \sum_{n=1}^{\infty} \frac{c_n J_0(a_n r/R_{\mathrm{T}}) \exp\left(-a_n^2 y\right)}{[1 + (a_n/B)^2] J_1^2(a_n)} \tag{5.13}$$

where

$$c_n = \frac{1}{R_{\mathrm{T}}^2} \int\limits_0^{R_{\mathrm{T}}} \left(\frac{T_{\mathrm{w}} - T_0(r)}{T_{\mathrm{w}} - T_0(0)}\right) J_0(a_n r/R_{\mathrm{T}}) r \, dr. \tag{5.13a}$$

5.1.4   Solution when Axial Heat Conduction is Considered

When flow rates are low, the axial heat conduction term cannot be ignored. Taking into account the axial heat conduction, Eq. (5.1) is solved for a semi-infinite bed. Under the boundary conditions expressed by

Eq. (5.3), the solution is

$$\frac{T_{\mathrm{w}} - T}{T_{\mathrm{w}} - T_0} = 2 \sum_{n=1}^{\infty} \frac{J_0(a_n r / R_{\mathrm{T}}) \exp(-a_n^2 z)}{a_n [1 + (a_n/B)^2] J_1(a_n)} \tag{5.14}$$

where

$$z = \frac{2y}{1 + \left[1 + 4\left(\dfrac{a_n}{GC_{\mathrm{F}}R_{\mathrm{T}}}\right)^2 k_{\mathrm{er}} k_{\mathrm{eax}}\right]^{1/2}} \tag{5.14a}$$

and y is defined in Eq. (5.4b).

When the inlet bed temperature is a function of r, and when axial heat conduction is taken into consideration, the solution of Eq. (5.1) for a semi-infinite bed is:

$$\frac{T_{\mathrm{w}} - T}{T_{\mathrm{w}} - T_0(0)} = 2 \sum_{n=1}^{\infty} \frac{c_n J_0(a_n r / R_{\mathrm{T}}) \exp(-a_n^2 z)}{[1 + (a_n/B)^2] J_1^2(a_n)} \tag{5.15}$$

where the coefficient c_n is defined in Eq. (5.13a).

At very low flow rates, Eq. (5.14a) reduces to

$$z = \frac{y}{\dfrac{a_n}{GC_{\mathrm{F}}R_{\mathrm{T}}} (k_{\mathrm{er}} k_{\mathrm{eax}})^{1/2}}. \tag{5.16}$$

As will be shown in Sections 5.2 and 5.5, the effective radial and axial thermal conductivities at low flow rates are equal to the effective thermal conductivity of a quiescent bed. Therefore, with Eq. (5.4b), Eq. (5.16) becomes

$$z = \frac{x}{a_n R_{\mathrm{T}}}. \tag{5.16a}$$

At sufficiently large values of x, only the first term of the series in Eqs.

(5.14) and (5.15) is dominant. Hence

$$\frac{T_w - T}{T_w - T_0} = \frac{2J_0(a_1 r/R_T) \exp(-a_1 x/R_T)}{a_1[1 + (a_1/B)^2] J_1(a_1)} \tag{5.17}$$

$$\frac{T_w - T}{T_w - T_0(0)} = \frac{2c_1 J_0(a_1 r/R_T) \exp(-a_1 x/R_T)}{[1 + (a_1/B)^2] J_1^2(a_1)}. \tag{5.18}$$

Equations (5.17) and (5.18) show that the bed temperature depends upon B, which is defined as $h_w R_T/k_{er}$, but not on the individual values of k_{er} and h_w.

5.1.5   Determination of Effective Radial Thermal Conductivities and Wall Heat Transfer Coefficients

The two heat transfer parameters, k_{er} and h_w, are easily determined from measurements of axial temperature profiles. The condition required is that the measurements should be made at high flow rates under which axial heat conduction may be ignored. The temperature profiles at any radial position will do, but the temperature measurements along the central axis of the bed, where radial temperatures level off, are most preferable. If the measurements are made near the wall, where radial temperature profiles are usually steep, small errors in radial location of the thermocouples will result in considerable errors in the measured temperatures.

If the measured temperatures at the center of the bed, T_c, are plotted as $\ln[(T_w - T_c)/(T_w - T_0)]$ versus x, a straight line will be produced at sufficiently large values of x. This straight line means that this is the region where Eq. (5.6) holds. The slope and intercept of the straight line are

$$\text{slope} = -a_1^2 \left(\frac{k_{er}}{GC_F R_T^2} \right) \tag{5.19}$$

and

$$\text{intercept} = \ln \frac{2}{a_1[1 + (a_1/B)^2] J_1(a_1)}. \tag{5.20}$$

Also, Eqs. (5.6) and (5.9) show that, when x is large, the mixed mean temperature of the effluent fluid, T_m, is related to T_c by

$$\frac{T_w - T_m}{T_w - T_c} = \frac{2J_1(a_1)}{a_1}. \tag{5.21}$$

Therefore, the parameters, k_{er} and h_w, can be determined by either (a) Eqs. (5.19) and (5.21), or (b) Eqs. (5.19) and (5.20), both in conjunction with Eqs. (5.4a) and (5.5a). In method (b), the value of a_1, based on Eqs. (5.20) and (5.5a), is easily effected by a slight change in the value of the intercept obtained from the extrapolation of the linear relationship between $\ln[(T_w - T_c)/(T_w - T_0)]$ and x. On the other hand, a_1 is more safely determined from Eq. (5.21). Therefore, k_{er} and h_w may, in general, be more accurately evaluated from method (a) than method (b).

Coberly and Marshall [2], however, evaluated the values of $\partial T/\partial x$, $\partial T/\partial r$ and $\partial^2 T/\partial r^2$ by graphical differentiation of the measured temperature profiles and then found k_{er} directly from Eq. (5.2). The local values of k_{er} were then averaged to give a mean value to be used for the entire bed. With this mean k_{er}, the value of a_1 was obtained from the slope of a straight line portion in a plot of $\ln[(T_w - T)/(T_w - T_0)]$ versus x. And finally h_w was calculated using Eqs. (5.4a) and (5.5a).

De Wasch and Froment [3] compared the measured temperature profiles at the bed exit and those computed from Eq. (5.4) and obtained k_{er} and h_w by adjusting them to give the best fit.

The various methods of determination of k_{er} and h_w from measured temperature profiles are illustrated in Examples 5.1 and 5.2.

Example 5.1

Table 5.1 lists the experimental conditions and the temperatures measured along the central axis of a cylindrical packed bed. Temperatures at the bed inlet were found to be almost constant in the radial direction. Find k_{er} and h_w.

SOLUTION

(i) *Estimation from Eqs. (5.19) and (5.21)*
The data are plotted as $\ln[(T_w - T_c)/(T_w - T_0)]$ versus x in Figure 5.2. The straight line portion has a slope of -0.064 cm^{-1}. With T_m and T_c at the bed exit, a_1 is found from Eq. (5.21) to be 1.63. Based on the slope and a_1,

TABLE 5.1
Experimental conditions and axial temperatures measured along
a bed central axis.

Bed diameter: 3.6 cm
Bed height: 20 cm
Glass bead: $D_p = 6.0$ mm
Air mass velocity: $G = 0.761$ kg m^{-2} s^{-1}
Specific heat of air: $C_F = 1000$ J kg^{-1} K^{-1}
Inlet air temperature: $T_0 = 16.5°C$
Wall temperature: $T_w = 100°C$

Height, x (cm)	Temperature at central axis, T_c (°C)
0	16.9
4	21.9
8	35.3
12	49.3
16	61.1
20	69.8

Mixed mean temperature of effluent air: 78.8°C

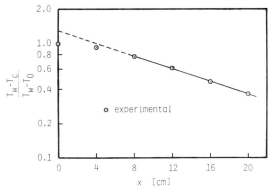

FIGURE 5.2 $\ln\left[(T_w - T_c)/(T_w - T_0)\right]$ versus x for Example 5.1.

k_{er} is determined from Eq. (5.19) to be 0.59 W m^{-1} K^{-1}, and then, according to Eqs. (5.4a) and (5.5a), $h_w = 70$ W m^{-2} K^{-1}.

(ii) *Estimation from Eqs. (5.19) and (5.20)*
In Figure 5.2, the extrapolation of the straight line portion to the y-axis gives an intercept equal to ln 1.24. From Eqs. (5.5a) and (5.20),

$$1.24 = \frac{2}{a_1\{1 + [J_0(a_1)/J_1(a_1)]^2\} J_1(a_1)}$$

and it is found that $a_1 = 1.35$. Also, from Eq. (5.5a), $B = 1.21$. From both the slope of the straight line, -0.064 cm^{-1}, and a_1, k_{er} can be determined from Eq. (5.19) to be 0.87 W m^{-1} K^{-1}. Hence, from Eq. (5.4a), $h_w = 58$ W m^{-2} K^{-1}.

(iii) *Estimation of k_{er} and h_w from curve fitting*

Temperatures along the central axis of the bed are computed from Eq. (5.4) with various assumed values of k_{er} and h_w, and then compared with the measured temperatures. The errors between the computed and measured temperatures, defined by Eq. (5.22),

$$\epsilon = \left\{ \frac{1}{N} \sum_{n=1}^{N} \left[\frac{(T_{expt} - T_{calc})n}{T_w - T_0} \right]^2 \right\}^{1/2} \tag{5.22}$$

are shown on an error map in Figure 5.3. (N is the number of points of comparison; in this case $N = 5$, corresponding to $x = 4$ to 20 cm.)

From the least-error point on the map (labeled +), the following data are obtained: $k_{er} = 0.71$ W m^{-1} K^{-1} and $h_w = 63$ W m^{-2} K^{-1}. Also, the data obtained in Solutions (i) and (ii) are indicated as $\odot$ and $\times$ on the map, respectively. As shown, the results obtained from Solution (i) are more accurate (with less error) than those of Solution (ii). For reference, the axial temperature profiles predicted with the data of k_{er} and h_w obtained in Solution (iii) are shown in Figure 5.4. It is shown that the first term in the series of Eq. (5.4) becomes dominant only when $x > 7$ cm.

Example 5.2

The experimental conditions and the measured temperatures listed in Table 5.2 are those of Bunnell *et al.* [4]. Find k_{er} and h_w. Also, examine the effect of k_{eax}.

SOLUTION

With an assumed value of $k_{eax} = 0.7$ W m^{-1} K^{-1}, bed temperatures are computed from Eq. (5.15). The predicted temperatures are then compared with the measured data. Figure 5.5(a) shows the resulting error map;

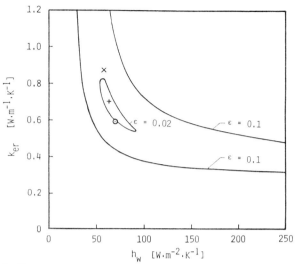

FIGURE 5.3 Error map in the plot of k_{er} versus h_w for Example 5.1 (Solution iii):
+ indicates the least-error point; ⊙ represents the data obtained in Solution (i);
× shows those in Solution (ii).

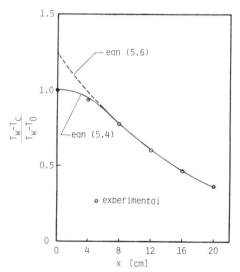

FIGURE 5.4 Comparison of measured and calculated temperature profiles for
Example 5.1.

TABLE 5.2

Experimental conditions and data measured by Bunnell et al. [4].

Bed diameter: 5.08 cm (2 in)
Cylindrical alumina pellet: $D_p = 0.32$ cm ($\frac{1}{8}$ in)
Air mass velocity: $G = 0.199$ kg m^{-2} s^{-1} (147 lb ft^{-2} h^{-1})
Specific heat of air: $C_F = 1050$ J kg^{-1} K^{-1}
Wall temperature: $T_w = 100°C$

x cm (in)	$r/R_T =$ 0	0.1	0.2	0.3	0.4	0.5	0.6	0.7	0.8	0.9
					Temperature (°C)					
0 (0)	400[a]	398[a]	396[a]	392[a]	388[a]	379[a]	367[a]	349[a]	320[a]	225[a]
5.1 (2)	318[a]		303		280		241		191	
10.2 (4)	234[a]		229		213		186		150	
15.2 (6)	182[a]		180		171		152		129	
20.3 (8)	157[a]		154		146		134		120	

[a] Read from their Figures 4–6.

ϵ being given by Eq. (5.22) with $N = 20$, corresponding to $x = 5.1$ to 20.3 cm. The least-error point (labeled +) indicates that $k_{er} = 0.31$ W m^{-1} K^{-1} and $h_w = 113$ W m^{-2} K^{-1}. Figures 5.5(b) and (c) show the effect of k_{eax}:

Figure 5.5(b) k_{er} versus k_{eax} with $h_w = 113$ W m^{-2} K^{-1};

Figure 5.5(c) h_w versus k_{eax} with $k_{er} = 0.31$ W m^{-1} K^{-1}.

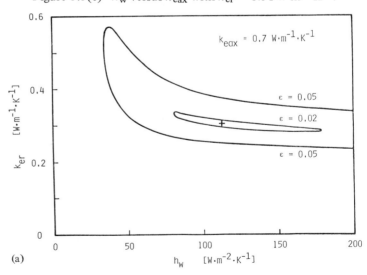

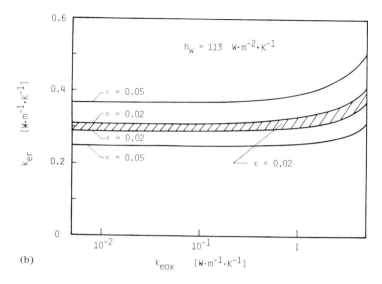

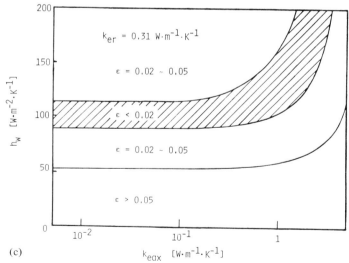

FIGURE 5.5 Error maps for the data of Bunnell *et al.* [4]: (a) k_{er} versus h_w (+ indicates the least-error point); (b) k_{er} versus k_{eax}; (c) h_w versus k_{eax}.

Bunnell *et al.* [4] found that their data of k_{er} were correlated by

$$\frac{k_{er}}{k_F} = 5.0 + 0.061 Re. \qquad (5.23)$$

From Eqs. (5.23) and (5.69), k_{eax} at this flow rate ($Re \doteqdot 30$, according to Bunnell *et al.*) is estimated to be roughly 0.7 W m^{-1} K^{-1}. However, Figures 5.5(b) and (c) indicate that, if $k_{eax} \lesssim 1$ W m^{-1} K^{-1}, axial heat conduction makes little contribution to the bed temperatures.

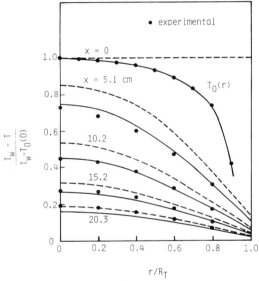

FIGURE 5.6   Measured and calculated radial temperature profiles (solid curves calculated from Eq. 5.15; dashed curves from Eq. 5.14).

Figure 5.6 shows the temperatures (solid lines) predicted from Eq. (5.15) with the measured $T_0(r)$, and those (dashed lines) computed from Eq. (5.14) assuming that the bed inlet temperature is uniform at $T_0(r = 0)$ = 400°C. The temperatures measured at $x = 5.1$, 10.2 and 15.2 cm show better agreement with the solid lines than with the dashed lines. It is clear, in this example, that the radially varied temperatures at the bed inlet should be taken into consideration.

(End of Example)

5.2   Effective Radial Thermal Conductivities

Hatta and Maeda [1, 5] correlated their data of effective radial thermal conductivities into the form: $k_{er}/k_F = a'Re^{b'}$ and obtained a' and b' as functions of the Reynolds number. Coberly and Marshall [2], however, found that their data were best fitted using a linear function of the Reynolds number as follows:

$$\frac{k_{er}}{k_F} = a + bRe \qquad (5.24)$$

where

> D_p = particle diameter
> k_F = fluid thermal conductivity
> $Re = D_p u \rho_F / \mu$, Reynolds number
> u = superficial fluid velocity, based on an empty column
> μ = fluid viscosity
> ρ_F = fluid density.

In 1952, Ranz [6] proposed a model for radial fluid mixing. His model is primarily based on the assumption that the bed is composed of inter-connected cells and that lateral mixing occurs from a single cell to the neighboring cells through randomly connected channels. The model further assumes that the average lateral fluid velocity is equal to a certain fraction of the apparent axial fluid velocity, u, i.e. αu. The lateral heat flow rate per unit area, through N interconnecting cells from temperatures T_0 to T_N in the bed, is then

$$
\begin{aligned}
q_{lateral} &= \alpha u \rho_F C_F (T_0 - T_1) \\
&= \alpha u \rho_F C_F (T_1 - T_2) \\
&= \ldots \\
&= \alpha u \rho_F C_F (T_{N-1} - T_N) \\
&= \frac{\alpha u \rho_F C_F}{N} (T_0 - T_N).
\end{aligned} \qquad (5.25)
$$

In terms of the effective radial thermal conductivity, Eq. (5.25) is re-written as:

$$q_{\text{lateral}} = (k_{er})_{\text{mixing}} \frac{T_0 - T_N}{Nl} \qquad (5.26)$$

where l is the distance between the centers of two adjacent cells or the width of a single cell. Therefore,

$$(k_{er})_{\text{mixing}} = \alpha u \rho_F C_F l. \qquad (5.27)$$

From a theoretical consideration of a high flow rate in a rhombohedral packing of spheres, α is found to be 0.179. In addition, Ranz assumes that $l = D_p/2$. Consequently,

$$\frac{(k_{er})_{\text{mixing}}}{k_F} = 0.0895 \,(Pr)(Re) \qquad (5.28)$$

where Pr is the Prandtl number, $C_F \mu/k_F$.

Yagi and Kunii [7, 8] proposed the following equation for the prediction of k_{er}:

$$\frac{k_{er}}{k_F} = \frac{k_e^0}{k_F} + \alpha\beta(Pr)(Re) \qquad (5.29)$$

where β comes from their assumption that $l = \beta D_p/2$. They found $\alpha\beta = 0.1$ from the data reported by Hatta and Maeda [1, 5] for cylindrical particles in the range of particle-to-column diameter ratio of 0.036 to 0.24. Also, they proposed a formula for estimating the values of k_e^0, the effective thermal conductivity of a quiescent bed, which is equivalent to a in Eq. (5.24).

Actually, the value of $\alpha\beta$ in Eq. (5.29) can be obtained based on a simple mass and heat transfer analogy. Turbulent fluid dispersion in the radial direction has already been found to be $(Pe_r)_{\text{mixing}} \simeq 10$ or $\epsilon_b(D_r)_{\text{mixing}} \simeq 0.1 D_p u$. Since the turbulent radial fluid dispersion coefficient, $\epsilon_b(D_r)_{\text{mixing}}$, and the effective radial thermal conductivity, $(k_{er})_{\text{mixing}}$, are both based on the unit cross-sectional area of a packed bed, and result from the lateral fluid mixing in the bed; then, $(k_{er})_{\text{mixing}} \simeq$

$0.1D_puC_F\rho_F$, or $\alpha\beta = 0.1$. Hence,

$$\frac{k_{er}}{k_F} = \frac{k_e^o}{k_F} + 0.1(Pr)(Re).$$ (5.30)

Similar to Eq. (2.25), it may be shown that

$$k_{er} = k_e^o \qquad\qquad \text{for laminar flow } (Re < 1)$$
$$= k_e^o + 0.1D_puC_F\rho_F \qquad \text{for turbulent flow } (Re > 5).$$ (5.30a)

However, the term, $0.1D_puC_F\rho_F$, is usually small compared to k_e^o at low flow rates, such that Eq. (5.30) may be applied over the entire range of flow from laminar to turbulent.

5.2.1   Effective Thermal Conductivities of Quiescent Beds

Many theoretical studies have been carried out on the estimation of effective thermal conductivities of quiescent beds. As listed in Table 5.3, the studies fall into two main groups: one assuming unidirectional heat flow, and the other considering two-dimensional heat flow. Unidirectional heat conduction was assumed by Schumann and Voss [9] for a hyperbolic solid; by Krischer and Kröll [12] for slab-array; and by Schlünder [13] for spheres. Also, the unidirectional heat flow of combined conduction and radiation was assumed by Argo and Smith [10], and Schotte [11] for spheres; by Yagi and Kunii [7, 8] for slab-array; and by Zehner and Schlünder [14] for spheroid solids.

The two-dimensional heat flow model is obviously more realistic than the unidirectional model. Deissler and Boegli [15] numerically solved the Laplace conduction equation by the relaxation method for a unit cell of spheres in a cubical array. The computed isotherms, in the cell with solid-to-fluid thermal conductivity ratios of $k_S/k_F = 3$ and 30, are shown in Figure 5.7. The isotherms come closer together in the vicinity of the point of contact as the ratio k_S/k_F increases from 3 to 30. It is clear that the heat flow is not unidirectional.

By considering that a packed bed may consist of a bundle of long cylinders as shown in Figure 5.8, Krupiczka [16, 17] solved numerically

TABLE 5.3

Theoretical studies of k_e^o. (Partly adapted from Zehner and Schlünder [14].)

Heat flow	Year	Investigator	Model	Solid	Heat transfer	Result
Unidirectional	1934	Schumann and Voss [9]		Hyperbolic $xy = m(m+1)$	Conduction	$\dfrac{k_e^o}{k_F} = \epsilon_b^3 + \dfrac{1-\epsilon_b^3}{m\left(\frac{k_F}{k_S}-1\right)+\frac{k_F}{k_S}}$ $\times\left[1+\dfrac{m(1+m)\left(\frac{k_F}{k_S}-1\right)\ln\left(\frac{k_F(1+m)}{k_S m}\right)}{m\left(\frac{k_F}{k_S}-1\right)+\frac{k_F}{k_S}}\right]$ $\epsilon_b = m(1+m)\ln\left(1+\dfrac{1}{m}\right)-m$
	1953	Argo and Smith [10]		Sphere	Conduction and radiation	Equation for k_e^o
	1960	Schotte [11]		Sphere	Conduction and radiation	Equation for k_e^o
	1954	Yagi and Kunii [7]		Slab	Conduction and radiation	$\dfrac{k_e^o}{k_F} = \epsilon_b\left(1+0.895\dfrac{h_{rv}D_p}{k_F}\right)$ $+\dfrac{0.895(1-\epsilon_b)}{\frac{1}{\phi+h_{rs}D_p/k_F}+\frac{2k_F}{3k_S}}$ $\phi = \text{func}\left(\dfrac{k_S}{k_F}\right)$ $h_{rs} = \dfrac{0.2268}{\frac{2}{p}-1}\left(\dfrac{T}{100}\right)^3 \text{ W m}^{-2}\text{K}^{-1}$ $h_{rv} = \dfrac{0.2268}{1+\frac{\epsilon_b}{2(1-\epsilon_b)}\frac{1-p}{p}}\left(\dfrac{T}{100}\right)^3 \text{ W m}^{-2}\text{K}^{-1}$

1956	Krischer and Kröll [12]		Slab	Conduction	$\dfrac{k_I}{k_F} = \epsilon_b + (1-\epsilon_b)\dfrac{k_S}{k_F}$ $\dfrac{k_{II}}{k_F} = \dfrac{1}{\epsilon_b + \dfrac{1-\epsilon_b}{k_S/k_F}}$ $\dfrac{k_e^o}{k_F} = \dfrac{1}{\dfrac{1-l}{k_I/k_F} + \dfrac{l}{k_{II}/k_F}}$
1966	Schlünder [13]		Sphere	Conduction	Graph for k_e^o
1970	Zehner and Schlünder [14]		Sphere $r^2 + \dfrac{z^2}{[B-(B-1)z]^2} = 1$ $B = 1$, sphere $B = \infty$, cylinder	Conduction and radiation	$\dfrac{k_e^o}{k_F} = \dfrac{2}{1-\dfrac{k_F}{k_S}}\left[\dfrac{\ln(k_S/k_F)}{1-\dfrac{k_F}{k_S}} - 1\right]$
Two-dimensional					
1958	Deissler and Boegli [15]		Sphere	Conduction	Graph for k_e^o
1966	Krupiczka [16]		Cylinder	Conduction	Eq. (5.32) for k_e^o
1969	Wakao and Kato [18]		Sphere	Conduction and radiation	Figure 5.9 for k_e^o
1971	Wakao and Vortmeyer [19]		Sphere	Conduction and radiation for fine particles and/or at low pressure	Graphs for k_e^o

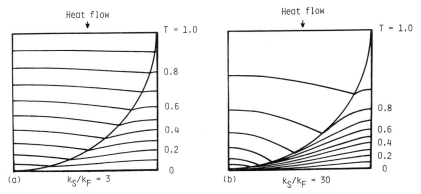

FIGURE 5.7 Isotherms in a cubic lattice of spheres, after Deissler and Boegli [15].

the following heat balance equations:

for a solid
$$\frac{1}{r}\frac{\partial}{\partial r}\left(r\frac{\partial T_S}{\partial r}\right)+\frac{1}{r^2}\frac{\partial^2 T_S}{\partial \theta^2}=0 \qquad 0<r<R$$

for a fluid
$$\frac{1}{r}\frac{\partial}{\partial r}\left(r\frac{\partial T_F}{\partial r}\right)+\frac{1}{r^2}\frac{\partial^2 T_F}{\partial \theta^2}=0 \qquad r>R$$

(5.31)

with the boundary conditions

temperature of ad = 1

temperature of bc = 0

ab and dc are adiabatic

$$T_S = T_F$$

$$k_S\frac{\partial T_S}{\partial r}=k_F\frac{\partial T_F}{\partial r} \qquad \text{at } r=R.$$

From the computed temperatures, the effective thermal conductivities of the quiescent cylinder beds ($\epsilon_b = 0.215$) are found to be expressed by

$$\log_{10}\frac{k_e^o}{k_F}=\left(0.785-0.057\log_{10}\frac{k_S}{k_F}\right)\log_{10}\frac{k_S}{k_F}. \qquad (5.32)$$

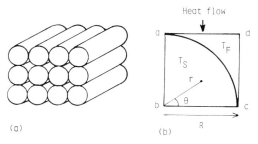

(a)   (b)

FIGURE 5.8   Model composed of cylinders, after Krupiczka [16, 17]: (a) model
composed of cylinders, $\epsilon_b = 0.215$; (b) cross-section.

Extending the results to a spherical lattice, Krupiczka proposed the
following equation for effective thermal conductivities of quiescent beds
of spherical particles:

$$\frac{k_e^o}{k_F} = \left(\frac{k_S}{k_F}\right)^n$$ (5.33)

with

$$n = 0.280 - 0.757 \log_{10} \epsilon_b - 0.057 \log_{10}\left(\frac{k_S}{k_F}\right).$$

When the fluid in the bed is gas, heat transfer by radiation becomes
significant at high temperatures. Under such conditions, combined conduction and radiation should then be considered as the mechanism of heat
transfer from the surface of a particle to surfaces of neighboring particles.
Wakao and Kato [18] calculated steady-state temperatures in a unit cell of
spheres arranged in an orthorhombic lattice with $\epsilon_b = 0.395$, taking
radiant heat transfer into account. The temperature computations were
made by the relaxation method for a network of grids. The effective
thermal conductivities, k_e^o, were then evaluated from the computed
temperature profiles. Figure 5.9 shows the values of k_e^o in a plot of k_e^o/k_F
versus k_S/k_F.

Moreover, Wakao and Kato [18] found that the values of k_e^o were
approximately expressed by Eq. (5.34)

$$k_e^o = (k_e^o)_{COND} + (k_e^o)_{RAD\text{-}COND}.$$ (5.34)

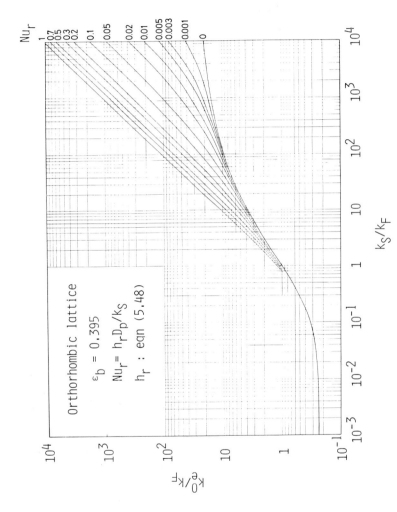

FIGURE 5.9   Static effective thermal conductivity for an orthorhombic lattice of spheres.

The first term is the effective thermal conductivity when heat transfer occurs only by conduction, which corresponds to the curve with a radiant Nusselt number, Nu_r, equal to zero in Figure 5.9. The second term shows the increase in effective thermal conductivity due to combined radiation and conduction at high temperatures. The values of $(k_e^0)_{RAD-COND}$ are approximately correlated by

$$\frac{(k_e^0)_{RAD-COND}}{k_F} = 0.707 Nu_r^{0.96} \left(\frac{k_S}{k_F}\right)^{1.11} \quad \text{for } 20 < \frac{k_S}{k_F} < 1000 \text{ and } Nu_r < 0.3$$

(5.35)

where Nu_r is defined as:

$$Nu_r = \frac{h_r D_p}{k_S}$$

(5.35a)

in which, h_r is the radiant heat transfer coefficient defined by Eq. (5.48) in the following section.

5.2.2 Radiant Heat Transfer Coefficients

Radiant heat flux between two large gray surfaces at temperatures T_1 and T_2 is given by:

$$q' = \frac{\sigma}{\frac{2}{p} - 1} (T_1^4 - T_2^4)$$

(5.36)

where σ, the Stefan–Boltzmann constant, is $5.67 \times 10^{-8} \text{ W m}^{-2} \text{ K}^{-4}$, and p is the emissivity of the gray surfaces.

The radiant heat flux may be written approximately in terms of an average temperature, T, as:

$$q' = \frac{4\sigma T^3}{\frac{2}{p} - 1} (T_1 - T_2).$$

(5.37)

Equation (5.37) is often expressed in the following form:

$$q' = h_r'(T_1 - T_2).$$

(5.38)

The radiant heat transfer coefficient, h'_r, is then

$$h'_r = \frac{4\sigma}{\dfrac{2}{p} - 1} T^3. \qquad (5.39)$$

This formula, first derived by Damköhler [20], was applied to heat transfer in packed beds by Argo and Smith [10], Yagi and Kunii [7, 8], and Kunii and Smith [21]. It should be mentioned that in packed beds the radiant heat flux, q', and the heat transfer coefficient, h'_r, are both defined per unit area of bed cross-section.

As well as Eq. (5.39), other formulae have also been used by different investigators. The one employed by Schotte [11] is

$$h'_r = 4\sigma p T^3. \qquad (5.40)$$

Chen and Churchill [22] derived Eq. (5.41) by applying the radiant two-flux model proposed by Hamaker [23]:

$$h'_r = 4\sigma \frac{2}{(a + 2b) D_p} T^3 \qquad (5.41)$$

where D_p is the particle diameter, a = absorption cross-section, and b = back scattering cross-section, both, defined per unit solid volume.

Vortmeyer [24] also applied the two-flux model and obtained the following formula:

$$h'_r = \frac{2B + p(1 - B)}{(2 - p)(1 - B)} 4\sigma T^3 \qquad (5.42)$$

where B is a geometrical bed constant. The value of B was found to be 0.1 by Vortmeyer and Börner [25].

Wakao and Kato [18] proposed a formula, which includes an overall view factor, for the evaluation of the radiant heat transfer coefficient based on the unit area of a particle surface. They assumed that every two hemispheres in contact were circumscribed with a diffusively reflective cylindrical wall, R_w, as shown in Figure 5.10. The overall view factor, $\bar{F}_{12}$, between the two hemispheres is defined as:

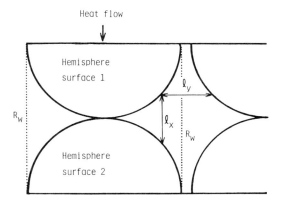

R$_w$: diffusively reflective wall

FIGURE 5.10   Two hemispheres in contact.

$$\bar{F}_{12} = F_{12} + \frac{F_{1R} F_{R2}}{1 - F_{RR}} \qquad (5.43)$$

where

F_{12} = view factor from surface 1 to surface 2
F_{1R} = view factor from surface 1 to wall R_w
F_{R2} = view factor from wall R_w to surface 2
F_{RR} = view factor of the wall R_w by itself.

For the cylindrical cell under consideration, the following relationships prevail:

$$\left.\begin{array}{l} F_{12} + F_{1R} = 1 \\[4pt] F_{RR} + F_{R1} + F_{R2} = 1 \\[4pt] F_{R1} = F_{R2}. \end{array}\right\} \qquad (5.44)$$

Moreover, the two view factors, F_{ij} and F_{ji}, are related reciprocally to their respective areas, A_i and A_j, by the following equation:

$$\frac{F_{ij}}{F_{ji}} = \frac{A_j}{A_i}.$$ (5.45)

The view factor between the two hemispheres in contact, F_{12}, has been computed to be 0.152. Hence, from Eqs. (5.43) to (5.45) the overall view factor, $\bar{F}_{12}$, is shown to be 0.576.

The radiant heat transfer rate, Q, between the two hemispheres at temperatures T_1 and T_2, respectively, is

$$Q = \frac{A\sigma}{2\left(\dfrac{1}{p} - 1\right) + \dfrac{1}{\bar{F}_{12}}}(T_1^4 - T_2^4)$$ (5.46)

where A is the surface area of the hemisphere. In terms of the radiant heat transfer coefficient, h_r, based on the unit area of a particle surface, the radiant heat transfer rate equation is written as:

$$Q = Ah_r(T_1 - T_2)$$ (5.47)

$$h_r = \frac{4\sigma}{2\left(\dfrac{1}{p} - 1\right) + \dfrac{1}{\bar{F}_{12}}} T^3$$

$$= \frac{0.2268}{\dfrac{2}{p} - 0.264}\left(\frac{T}{100}\right)^3 \quad (\text{W m}^{-2}\,\text{K}^{-1}).$$ (5.48)

It should be noted that the definitions of h_r and h_r' are based on different unit areas. In the cell shown in Figure 5.10, if the cross-sectional area of the cell is denoted by A', the two heat transfer coefficients are related in the following manner:

$$h_r A = h_r' A'.$$ (5.49)

The area of the hemisphere is roughly twice the cross-sectional area of the cell, i.e. $A \simeq 2A'$, therefore,

$$h_r = \frac{h'_r}{2}. \tag{5.50}$$

These formulae are compared in terms of $h'_r/(4\sigma T^3)$ in Table 5.4. It is found that the values of radiant heat transfer coefficients predicted from Eq. (5.42) by Vortmeyer [24] and Eq. (5.48) by Wakao and Kato [18] are in fairly good agreement.

TABLE 5.4
Comparison of radiant heat transfer coefficients.

	$h'_r/(4\sigma T^3)$			
Emissivity, p	Damköhler [20] Eq. (5.39)	Schotte [11] Eq. (5.40)	Vortmeyer [24] Eq. (5.42) with $B = 0.1$	Wakao and Kato [18] Eq. (5.48)
0.2	0.11	0.2	0.23	0.21
0.5	0.33	0.5	0.48	0.54
0.8	0.67	0.8	0.85	0.89
1	1	1	1.2	1.2

5.2.3 Effect of Gas Radiation

In the case of the void space in the cell of Figure 5.10 being filled with a radiating gray gas, Wakao [26] has shown that the radiant heat transfer coefficient given by Eq. (5.48) can be corrected by the introduction of emissivity of the gray gas, p_g, as follows:

$$h^*_r = \frac{4\sigma}{2\left(\frac{1}{p} - 1\right) + \dfrac{2}{1 + F_{12}(1 - p_g)}} T^3. \tag{5.51}$$

The gas emissivity is a function of gas temperature, size of the void and the partial pressures of the radiating gases in the void. In packed beds, the size of the void is usually small so that the gas emissivity is very low. For example, the emissivity of pure CO_2 at a temperature of as high as 900°C in a large void of 1 cm in an average beam length is found [27] to be only 0.05.

Because of the low gas emissivities, the effect of radiating gas on the radiant heat transfer coefficient in a packed bed is usually insignificant. In other words, gas radiation has little effect on the effective thermal conductivity of packed bed. Equation (5.48) can be applied, therefore, even for radiant heat transfer in packed beds containing radiating gases.

5.2.4   Effective Thermal Conductivities of Quiescent Beds at Low Pressures

The effective thermal conductivities of quiescent beds shown in Figure 5.9 are those computed assuming that fluid thermal conductivity is kept constant and does not depend on pressure. This is true only under normal pressure conditions where the length of the mean free path of the gas molecule is considerably less than the characteristic lengths of the interstitial volumes. However, there are always some regions near the contact points of the particles where this condition is not fulfilled. Under normal pressure conditions heat transfer in these regions can be neglected since the area of these regions is very small compared with the total heat transfer area. The effective thermal conductivities of quiescent beds for particles larger than about 1 mm, under normal pressure conditions, can be predicted from Figure 5.9.

In beds with particle diameters of a few millimeters, the effective thermal conductivity is constant at normal pressure, but decreases with decreasing pressure when the pressure is sufficiently low. This decrease in effective thermal conductivity is due to a temperature jump at the particle surface. For fine powder particles, however, this effect of decreasing effective thermal conductivity usually occurs at atmospheric or even higher pressures.

The temperature jump effect is well known from the kinetic theory of gases. One of the first investigations was made by Smoluchowsky [28] who proposed an equation, which was later modified by Hengst [29], of the following form:

$$k_e^o = b \cfrac{\cfrac{\pi}{2} k_F}{1 - \cfrac{k_F}{k_S}} \left[1 + \cfrac{2\delta}{D_p} \ln \left(\cfrac{1 + \cfrac{2\delta}{D_p}}{\cfrac{k_F}{k_S} + \cfrac{2\delta}{D_p}} \right) - 1 \right] + k_F \left(1 - \frac{\pi}{4} \right) + a \quad (5.52)$$

where k_F is the gas thermal conductivity under normal pressure, δ is the temperature jump coefficient defined as: $\delta = c/P$ (c is an experimental constant and P is the pressure), a is the effective thermal conductivity in a vacuum, and b is the correction factor for beds compared with a cubic lattice of spheres.

Kling [30] measured the effective thermal conductivities for steel shot–hydrogen/air systems. He applied the above formula to the measured data and by choosing suitable values for the parameters b and c was able to obtain good agreement between the predicted and theoretical results.

In addition to the above, measurements of effective thermal conductivities in various types of beds have been reported: Fulk [31] for beds of perlite particles at air pressure from 0.1 MPa to 0.1 Pa; Kaganer and Glebova [32] for silica gels at air pressure from 0.1 MPa to 1 Pa; Masamune and Smith [33] for beds of glass beads and steel shot at air pressure from 0.1 MPa to 1 Pa; Swift [34] for fine particles of uranium, zirconium and oxidized uranium in helium, nitrogen, argon and methane at pressures from 0.1 MPa to 1 or 1×10^{-3} Pa; Wakao et al. [35] for glass beads in helium, hydrogen, nitrogen, acetylene and ethylene and lead shot in hydrogen in the pressure range 0.1 MPa to 0.1 Pa; and Luikov et al. [36] for quartz sand and powdered plexiglass at air pressure from 0.1 MPa to 0.1 Pa. The results obtained by these investigators all show the same characteristic S-shaped curves when the effective thermal conductivities are plotted against pressure using logarithmic coordinates.

5.2.4.1   Heat transfer between two parallel plates at low pressure

As mentioned already, the kinetic gas theory predicts a temperature jump at the surface. The temperature jump leads to a decrease in the gas conductivity, and consequently, to a decrease in the effective thermal conductivity of packed beds.

Following some simple arguments by Chapman and Cowling [37], the temperature jump, in the case of two parallel plates under the condition that the length of the mean free path, λ, is less than the distance between the two plates, l, may be described by

$$\Delta T = \left(\frac{2 - \alpha}{\alpha}\right) \lambda \left(\frac{dT}{dx}\right)_{inner} \tag{5.53}$$

where α is the accommodation coefficient, and $(dT/dx)_{inner}$ is the temperature gradient in the range where the gas can still be considered as a continuum.

With the further assumption that the temperature jump occurs along one mean free path length, the thermal conductivity of gas is given by Eq. (5.54).

$$k_F^* = \frac{k_F}{1 + \frac{2(2-\alpha)}{\alpha}\frac{\lambda}{l}}$$

$$= \frac{k_F}{1 + \frac{2(2-\alpha)}{\alpha}\frac{\lambda_0 P_0}{lP}} \tag{5.54}$$

where P_0 is the atmospheric pressure and λ_0 is the length of the mean free path at P_0. The values of λ_0 at 0 and 20°C are listed for some gases in Table 5.5.

When the pressure is further decreased to a range where $\lambda \sim l$, the assumptions leading to Eq. (5.53) are no longer valid. Now the molecules just hit the surfaces without colliding with each other. From the kinetic theory of gases, the rate of heat transfer in this range should be exactly proportional to the pressure. Since Eq. (5.54) also shows this linear

TABLE 5.5
Mean free paths of several gases at atmospheric pressure.

Gas	Mean free path (nm)	
	0°C	20°C
Ar	88.6	97.5
CO	83.5	91.1
CO_2	54.9	60.7
H_2	157.9	172.1
He	249.2	270.9
Kr	93.8	
N_2	83.9	91.7
NH_3	58.4	65.1
O_2	89.3	98.0
Xe	55.3	

The Boltzmann mean free paths evaluated from the data at 0.100 MPa in the *Handbook of Chemistry and Physics* [38].

pressure dependency, the equation may be used to predict k_F^* for very low pressure ranges. In fact, the error associated with the application of Eq. (5.54) in the case of $\lambda \sim l$, is not as serious as one might think; since in packed beds within this pressure range, solid–solid conduction and radiation contributions are dominant.

5.2.4.2 Calculation of the effective thermal conductivities of quiescent beds

Wakao and Vortmeyer [19] proposed a method for the computation of effective thermal conductivities of quiescent beds of fine powder, taking the effect of contact conductivity into account. The calculation procedure is identical to that employed by Wakao and Kato [18] except that k_F^* of Eq. (5.54) is used as the distance dependent gas thermal conductivity.

The conductivity of a grid in the direction of heat flow or the x-direction is

$$k_{Fx}^* = \frac{k_F}{1 + \dfrac{D_p}{l_x \Psi}} \tag{5.55}$$

and that in the direction perpendicular to the heat flow or the y-direction is

$$k_{Fy}^* = \frac{k_F}{1 + \dfrac{D_p}{l_y \Psi}} \tag{5.56}$$

where l_x and l_y are the distances (refer to Figure 5.10) between two surface elements on the x and y grids, respectively.

The pressure parameter, Ψ, is defined as:

$$\begin{aligned}
\Psi &= \frac{D_p}{\lambda} \frac{\alpha}{2(2-\alpha)} \\
&= \frac{D_p}{\lambda_0 P_0} \frac{\alpha}{2(2-\alpha)} P.
\end{aligned} \tag{5.57}$$

The steady-state temperatures at nodal points are computed by the relaxation method, and consequently, the effective thermal conductivities can be evaluated. Based on the calculated results, Wakao and Vortmeyer [19] suggested the following expression for the model:

$$k_e^o = (k_e^o)_{COND} + (k_e^o)_{RAD\text{-}COND} + (k_e^o)_{CONTACT}. \qquad (5.58)$$

The conduction term, $(k_e^o)_{COND}$, is shown as $(k_e^o)_{COND}/k_F$ versus k_S/k_F in Figure 5.11. It should again be noted that k_F is the gas thermal conductivity under normal pressure.

The combined radiation and conduction term, $(k_e^o)_{RAD\text{-}COND}$, is approximately expressed as:

$$\frac{(k_e^o)_{RAD\text{-}COND}}{k_S} = \frac{1.8Nu_r}{1 + 0.7Nu_r} f \qquad \text{for } Nu_r < 0.5. \qquad (5.59)$$

The term $1.8Nu_r/(1+0.7Nu_r)$ corresponds to $(k_e^o)_{RAD\text{-}COND}/k_S$ in a vacuum or at $\Psi = 0$. Figure 5.12 shows the correction factor, f, as a function of k_S/k_F and Ψ.

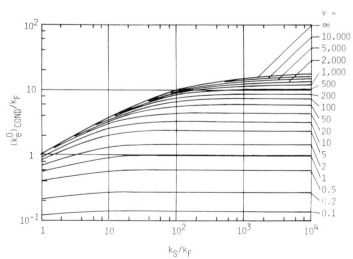

FIGURE 5.11 $(k_e^o)_{COND}/k_F$ versus k_S/k_F.

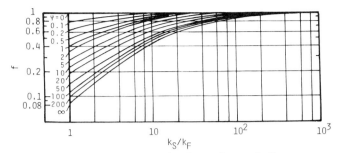

FIGURE 5.12   Correction factor f versus k_S/k_F.

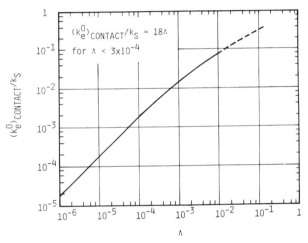

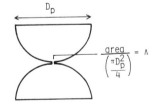

FIGURE 5.13 $(k_e^o)_{CONTACT}/k_S$ versus Λ.

The solid–solid contact term $(k_e^o)_{CONTACT}$ is very small under normal pressure conditions, but becomes increasingly significant with lowering pressure. Figure 5.13 is a plot of $(k_e^o)_{CONTACT}/k_S$ versus Λ, in which the

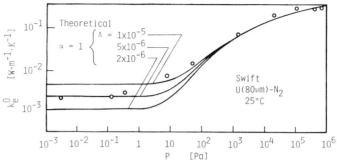

FIGURE 5.14 k_e^o versus pressure, for a uranium powder–nitrogen system.

fractional contact area, Λ, is the ratio of the solid–solid contact area to the projected area of a particle, $\pi D_p^2/4$. When Λ is small, $(k_e^o)_{CONTACT}$ is expressed as

$$\frac{(k_e^o)_{CONTACT}}{k_S} = 18\Lambda \qquad \text{for } \Lambda < 3 \times 10^{-4}. \qquad (5.60)$$

Among the published data of effective thermal conductivities of quiescent beds, Swift's [34] graphs are regarded as the most comprehensive. Thus, some of his data on fine particles of uranium and zirconium are compared with the model proposed by Wakao and Vortmeyer [19] as follows: Figure 5.14 demonstrates how the contact areas between two particles affect the asymptotic value of k_e^o at low pressure. The data in a vacuum consist of solid–solid contact and radiation conductivities. Because of the fine powder and low temperature, the radiation conductivity at $P < 1$ Pa is small: $[(k_e^o)_{RAD-COND}]_{\Psi=0} = 5 \times 10^{-4}$ W m^{-1} K^{-1}. In calculating the theoretical curves, the accommodation coefficient is assumed to be unity. Therefore, $(k_e^o)_{CONTACT} = 2.22 \times 10^{-3}$ W m^{-1} K^{-1} and consequently the contact area fraction for uranium powder ($k_S = 24.7$ W m^{-1}K^{-1}) is found to be $\Lambda = 5 \times 10^{-6}$.

The data, given in Figures 5.15(a)–(c), are for beds of zirconium powder ($k_S = 26$ W m^{-1} K^{-1}) filled with argon, nitrogen and helium, respectively. At a pressure of 1.3 Pa, the measured data are equal to each other. This consists of $(k_e^o)_{CONTACT}$ and $[(k_e^o)_{RAD-COND}]_{\Psi=0}$. The contact area fraction, Λ, is eventually found to be 7×10^{-6}. The graphs demonstrate

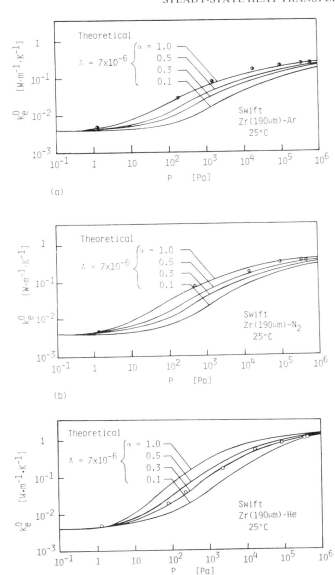

FIGURE 5.15 k_e^0 versus pressure: (a) zirconium powder–argon system; (b) zirconium powder–nitrogen system; (c) zirconium powder–helium system.

how the effective thermal conductivities are influenced by the values of the accommodation coefficient. It is found that the experimental data are best fitted by assuming $\alpha = 1$ for argon and nitrogen, and $\alpha = 0.3$ for helium. The low accommodation coefficients for lighter gases such as helium and hydrogen are in good agreement with the reported literature values [39–41]. It should be pointed out that the effective thermal conductivities for such fine particles are still influenced by the accommodation coefficient even at normal pressure.

The small contact area fraction, $\Lambda = (5 \sim 7) \times 10^{-6}$, as found from Figures 5.14 and 5.15, is due to the fact that a rather small apparatus (cylindrical bed: 5.08 cm (2 in) long and 1.9 cm (0.75 in) diameter) was employed in the measurements.

Kling [30] measured the effective thermal conductivities of beds of steel shot–hydrogen/air, and determined the values of the parameters b and c in Hengst's equation, Eq. (5.52), for the systems studied. The contributions of solid-solid contact and radiation to the effective thermal conductivities were not stated in their paper, therefore, only the gas–solid conduction contribution (the first and second terms on the right hand side of Eq. 5.52) is compared with the $(k_e^0)_{COND}$ term of Eq. (5.58). Note that, because of the high thermal conductivity of steel shot $(k_S/k_F = 200$

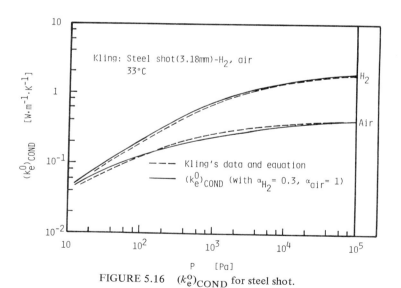

FIGURE 5.16 $(k_e^0)_{COND}$ for steel shot.

for hydrogen and $k_S/k_F = 1500$ for air), the correction factor, f, for radiation conductivity is almost unity. Therefore, $(k_e^o)_{RAD\text{-}COND}$ is independent of pressure and $(k_e^o)_{RAD\text{-}COND} + (k_e^o)_{CONTACT}$ corresponds to a in Eq. (5.52). Figure 5.16 shows good agreement between the Hengst–Kling equation and the model proposed by Wakao and Vortmeyer [19].

The contact conductivity is considered to depend on such physical properties as hardness, plasticity and thermal expansion of the particles, roughness of the particle surfaces, bed weight and external mechanical load. If the particle is covered with an oxide film, for instance, an additional resistance should be accounted for in $(k_e^o)_{CONTACT}$. At present, there is no way of predicting the value of $(k_e^o)_{CONTACT}$, and this should be evaluated only from effective thermal conductivity measurements in a vacuum.

5.3   Wall Heat Transfer Coefficients

Wall heat transfer coefficients, h_w, have also been measured by many research workers. From an examination of the published data in the literature Li and Finlayson [42] found that a large number of the data of h_w had a bed length effect or entrance effect: some of the h_w data

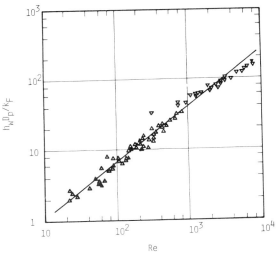

FIGURE 5.17 $h_w D_p/k_F$ versus Re for a spherical particle-air system, after Li and Finlayson [42].

were obtained from Eqs. (5.19) to (5.21) even though plots of the $\ln[(T_w - T)/(T_w - T_0)]$-$x$ relationships did not quite form a straight line.

5.3.1 Spherical Particle-air System

Based on the published h_w data of Yagi and Wakao [43] and Kunii *et al.* [44], which are considered to be free from the length effect, Li and Finlayson [42] plotted $h_w D_p/k_F$ versus Re as shown in Figure 5.17, and from which they obtained the following correlation:

$$\frac{h_w D_p}{k_F} = 0.17 Re^{0.79} \tag{5.61}$$

for $20 \leqslant Re \leqslant 7600$ and $0.05 \leqslant D_p/D_T \leqslant 0.3$, where D_T is the column diameter.

5.3.2 Cylindrical Particle-air System

Similarly, Li and Finlayson [42] correlated the selected h_w data obtained by Hatta and Maeda [1, 5], Coberly and Marshall [2], Felix [45], Phillips *et al.* [46], and Hashimoto *et al.* [47] as shown in Figure 5.18. They suggested the following empirical relationship between $h_w D_p/k_F$ and Re:

$$\frac{h_w D_p}{k_F} = 0.16 Re^{0.93} \tag{5.62}$$

for $20 \leqslant Re \leqslant 800$, $0.03 \leqslant D_p/D_T \leqslant 0.2$ and $D_p = 6V_p/S_p$, where S_p and V_p are the surface area and volume of a particle, respectively.

Yagi and Kunii [48] correlated their data of h_w into the following form:

$$\frac{h_w D_p}{k_F} = a' + 0.054(Pr)(Re) \tag{5.63}$$

for $20 \leqslant Re \leqslant 2000$, where a' depends on the shape of the solid particle, D_p and D_p/D_T. The a' values for glass beads and Celite cylinders were found to be:

a'	D_p (mm)	D_p/D_T
1.2	$0.8 \sim 0.9$	$0.02 \sim 0.04$
3	$1.8 \sim 3.2$	$0.04 \sim 0.07$
5	$4.3 \sim 6.4$	$0.08 \sim 0.17$

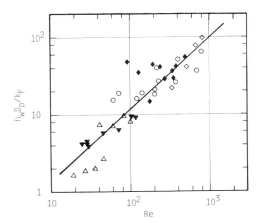

FIGURE 5.18 $h_w D_p / k_F$ versus Re for a cylindrical particle–air system, after Li and Finlayson [42].

However, Gunn and Khalid [49] found that their experimental data with metal and glass beads, for $Re = 2$ to 400, were larger than the values estimated from Eq. (5.61).

The heat balance equation, Eq. (5.1), is generally derived based on the assumption that the fluid mass velocity per unit area of bed cross-section is constant. However, Schwartz and Smith [50], in their work on the measurement of radial fluid velocities, found that the radial velocity profile, as shown in Figure 5.19, reached a maximum value at about one particle diameter away from the column wall. Their experimental results subsequently led to studies of radial voidage variations by several groups of researchers.

In the experimental work carried out by Kimura *et al.* [51] and Roblee *et al.* [52], the void space in the bed was filled with molten wax, which was then allowed to solidify. A slab of the bed composed of the particles and wax was mounted in a turning lathe chuck. Annular layers were shaved off and the void fraction of each layer was found from the volume of the melted wax collected for the layer. Kimura *et al.* measured the radial voidage variation for beds of broken pieces of limestone, and Roblee *et al.* did this for beds of cork spheres, wood cylinders, carbon Raschig rings and carbon Berl saddles. In the case of spheres, the voidage variation,

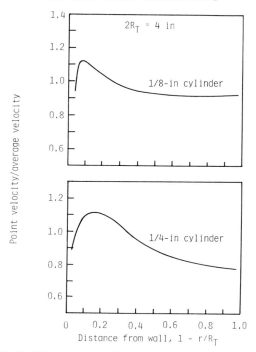

FIGURE 5.19   Radial velocity profiles measured by Schwartz and Smith [50], for $\frac{1}{8}$ in (0.32 cm) and $\frac{1}{4}$ in (0.64 cm) diameter cylinders in a 4 in (10 cm) diameter column.

shown in Figure 5.20, indicates that the voidage reaches a minimum at one particle radius from the wall, within alternate maximums and minimums occurring at successive particle radii. The amplitude of the cycling decreases as the distance from the wall increases.

Pillai [53], instead, counted the number of particles in the vicinity of the wall in a two-dimensional vessel. The results show, similar to those of Roblee *et al.* [52], that the voidage variation follows a heavily damped oscillatory curve with high voidage at about one particle diameter away from the wall. This explains well the experimental results of Schwartz and Smith [50] that velocity is at a maximum at about one particle diameter from the wall. The large voidage variation near the wall makes the velocity profile and effective thermal conductivity near the wall differ from those in the bed core. In solving Eqs. (5.1) and (5.2), however, both G and k_{er} are assumed to be constant. Hence, the wall heat transfer coefficient not

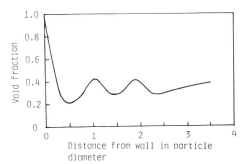

FIGURE 5.20 Radial voidage variation measured by Roblee *et al.* [52], for 0.76 in (1.9 cm) diameter spheres in a 6.7 in (17 cm) diameter column.

only stands for the intrinsic heat transfer at the wall, but also includes the effects resulting from radial variations in the fluid velocity and k_{er} in the vicinity of the wall.

5.4 Overall Heat Transfer Coefficients

The overall heat transfer coefficients, defined according to Eq. (5.10), have been determined experimentally for various solid–fluid systems. Empirical correlations obtained by Li and Finlayson [42] are presented as follows:

5.4.1 Spherical Particle-air System

$$\frac{U_0 D_T}{k_F} = 2.03 Re^{0.8} \exp\left(-\frac{6D_p}{D_T}\right) \tag{5.64}$$

for $20 \leqslant Re \leqslant 7600$ and $0.05 \leqslant D_p/D_T \leqslant 0.3$.

5.4.2 Cylindrical Particle-air System

$$\frac{U_0 D_T}{k_F} = 1.26 Re^{0.95} \exp\left(-\frac{6D_p}{D_T}\right) \tag{5.65}$$

for $20 \leqslant Re \leqslant 800$ and $0.03 \leqslant D_p/D_T \leqslant 0.2$.

5.5   Effective Axial Thermal Conductivities

Yagi *et al.* [54] were the first to obtain the effective axial thermal conductivities of packed beds. As shown in Figure 5.21, their axial steady-state heat transfer measurements were made with the bed packed in an adiabatic column. The packed bed was heated from the top by an infra-red lamp so that heat penetrated downwards into the bed, while air flowed countercurrently upwards through the bed from the bottom.

From the following heat balance equation:

$$GC_F \frac{dT}{dx} + k_{eax} \frac{d^2T}{dx^2} = 0 \qquad (5.66)$$

where x is the distance in the bed measured from the exit, the axial temperature is found to be

$$T - T_0 \propto \exp\left(-\frac{GC_F x}{k_{eax}}\right) \qquad (5.67)$$

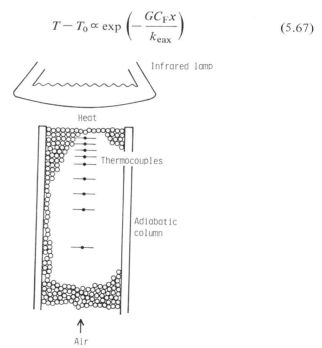

FIGURE 5.21   Apparatus used for effective axial thermal conductivity measurements by Yagi *et al.* [54].

where T_0 is the temperature of air flowing into the packed bed. Therefore, the values of k_{eax} were obtained from measurements of axial temperature profiles. The data were found to be correlated by

$$\frac{k_{eax}}{k_F} = \frac{k_e^o}{k_F} + \delta(Pr)(Re) \qquad (5.68)$$

with $\delta = 0.8$ for glass beads and limestone broken pieces, and $\delta = 0.7$ for metal spheres at low air flow rates.

By the time Yagi *et al*. had finished their work on the measurements of k_{eax} no research had been reported on effective axial thermal conductivity. Since they thought that k_{eax} should be lower than k_{er}, they were rather puzzled to find that the coefficient, δ, of the second term on the right hand side of Eq. (5.68) was larger than that of Eq. (5.30) for effective radial thermal conductivity, and hesitated to publish their results.

However, in the March 1957 issue of *AIChE J*., McHenry and Wilhelm [55] published the results of their work on mass dispersion in packed beds in which they had made frequency response measurements with the binary gas systems, H_2-N_2 and C_2H_4-N_2, in beds of non-porous particles. They found that $Pe_{ax} = 1.88 \pm 0.15$ in the range of Reynolds number, 10 to 400. Also, McHenry and Wilhelm came to realize, from a theoretical considera-tion, that an axial Peclet number of two should be expected if the frequency response measurements were made with a series of n perfect mixers, where n was the number of particles traversed between the bed inlet and outlet. Soon after this, in the June 1957 issue of *AIChE J*., Aris and Amundson [56] published a theoretical work in which they reached the same conclusion as that of McHenry and Wilhelm.

As far as beds of Raschig rings with flowing water are concerned, axial fluid dispersion coefficients were measured as early as 1953 by Danckwerts [57] and Kramers and Alberda [58]. Danckwerts obtained $Pe_{ax} = 1/1.8$ at $Re = 24$, and Kramers and Alberda found $Pe_{ax} = 1/(1.1 \pm 0.1)$ for $Re = 75$ to 150. Their Pe_{ax} results are lower than those of McHenry and Wilhelm [55]. In any case, the work of McHenry and Wilhelm made Yagi *et al*. [54] submit their results on the effective axial thermal conductivity for publica-tion, and it appeared in *AIChE J*., three years after the completion of the work.

The axial fluid mixing coefficient per unit cross-sectional area of a packed bed has been shown to be $0.5D_p u$. According to a heat–mass analogy, the contribution of turbulent flow to the effective axial thermal conductivity is considered to be $0.5 D_p u C_F \rho_F$. Therefore, Eq. (5.69) is recommended for the evaluation of k_{eax} in a wide range of Reynolds number.

$$\frac{k_{eax}}{k_F} = \frac{k_e^o}{k_F} + 0.5\,(Pr)(Re).$$ (5.69)

REFERENCES

[1]   S. Hatta and S. Maeda, *Kagaku Kogaku* **12**, 56 (1948).
[2]   C. A. Coberly and W. R. Marshall, *Chem. Eng. Prog.* **47**, 141 (1951).
[3]   A. P. De Wasch and G. F. Froment, *Chem. Eng. Sci.* **27**, 567 (1972).
[4]   D. G. Bunnell, H. B. Irvin, R. W. Olson and J. M. Smith, *IEC* **41**, 1977 (1949).
[5]   S. Hatta and S. Maeda, *Kagaku Kogaku* **13**, 79 (1949).
[6]   W. E. Ranz, *Chem. Eng. Prog.* **48**, 247 (1952).
[7]   S. Yagi and D. Kunii, *Kagaku Kogaku* **18**, 576 (1954).
[8]   S. Yagi and D. Kunii, *AIChE J.* **3**, 373 (1957).
[9]   T. E. W. Schumann and V. Voss, *Fuel* **13**, 249 (1934).
[10] W. B. Argo and J. M. Smith, *Chem. Eng. Prog.* **49**, 443 (1953).
[11] W. Schotte, *AIChE J.* **6**, 63 (1960).
[12] O. Krischer and K. Kröll, *Die wissenschaftlichen Grundlagen der Trocknungs-technik.*, Bd. 1, Berlin-Göttingen-Heiderberg (1956).
[13] E. U. Schlünder, *Chem. Ing. Tech.* **38**, 967 (1966).
[14] P. Zehner and E. U. Schlünder, *Chem. Ing. Tech.* **42**, 933 (1970).
[15] R. G. Deissler and J. S. Boegli, *Trans. ASME* **80**, 1417 (1958).
[16] R. Krupiczka, *Chemia Stosowana* **2B**, 183 (1966).
[17] R. Krupiczka, *Int. Chem. Eng.* **7**, 122 (1967).
[18] N. Wakao and K. Kato, *J. Chem. Eng. Japan* **2**, 24 (1969).
[19] N. Wakao and D. Vortmeyer, *Chem. Eng. Sci.* **26**, 1753 (1971).
[20] G. Damköhler, *Der Chemie Ingenieur, Eucken-Jacob*, Vol. 3, Akadem. Verlag. Leipzig, p. 445 (1937).
[21] D. Kunii and J. M. Smith, *AIChE J.* **6**, 71 (1960).
[22] J. C. Chen and S. W. Churchill, *AIChE J.* **9**, 35 (1963).
[23] H. C. Hamaker, *Philips Res. Reports* **2**, 55, 103, 112, 420 (1947).
[24] D. Vortmeyer, *Fortschr. Ber. VDI-Z* Reihe 3, Nr. 9, VDI-Verlag., Düsseldorf (1966).
[25] D. Vortmeyer and C. J. Börner, *Chem. Ing. Tech.* **38**, 1077 (1966).
[26] N. Wakao, *Chem. Eng. Sci.* **28**, 1117 (1973).
[27] H. C. Hottel, *Heat Transmission*, edited by W. C. McAdams, 3rd edn., McGraw-Hill, New York, Ch. 2 (1954).
[28] M. Smoluchowsky, *Wiener Akad.* **107**, 304 (1898).
[29] G. Hengst, *Dissertation* Technische Universität München (1934).
[30] G. Kling, *VDI-Forschung* **9**, 28 (1938).
[31] M. M. Fulk, *Progress in Cryogenics* **1**, 63 (1959).
[32] M. G. Kaganer and L. I. Glebova, *Kislorod* **1**, 13 (1959).
[33] S. Masamune and J. M. Smith, *Ind. Eng. Chem. Fund.* **2**, 136 (1963).
[34] D. L. Swift, *Int. J. Heat Mass Transfer* **9**, 1061 (1966).
[35] N. Wakao, S. Omura and M. Fukuda, *Kagaku Kogaku* **30**, 1119 (1966).
[36] A. V. Luikov, A. G. Shashkov, L. L. Vasiliev and Y. E. Fraiman, *Int. J. Heat Mass Transfer* **11**, 117 (1968).

[37]   S. Chapman and T. G. Cowling, *The Mathematical Theory of Non-Uniform Gases*, Cambridge University Press, p. 104 (1960).
[38]   *Handbook of Chemistry and Physics*, edited by R. C. Weast, 53rd edn., Chemical Rubber Company, Ohio, F-174 (1972–73).
[39]   M. Knudsen, *Ann. Phys.* **34**, 593 (1911).
[40]   J. H. Wachmann, *PhD thesis*, University of Missouri (1957).
[41]   M. L. Wiedmann and P. R. Trumpler, *Trans. ASME* **68**, 57 (1946).
[42]   C. H. Li and B. A. Finlayson, *Chem. Eng. Sci.* **32**, 1055 (1977).
[43]   S. Yagi and N. Wakao, *AIChE J.* **5**, 79 (1959).
[44]   D. Kunii, M. Suzuki and N. Ono, *J. Chem. Eng. Japan* **1**, 21 (1968).
[45]   T. R. Felix, *PhD thesis*, University of Wisconsin (1951).
[46]   B. D. Phillips, F. W. Leavitt and C. Y. Yoon, *Chem. Eng. Prog. Symp. Ser.* **56** (No. 30), 219 (1960).
[47]   K. Hashimoto, N. Suzuki, M. Teramoto and S. Nagata, *Kagaku Kogaku* **4**, 68 (1966).
[48]   S. Yagi and D. Kunii, *AIChE J.* **6**, 97 (1960).
[49]   D. J. Gunn and M. Khalid, *Chem. Eng. Sci.* **30**, 261 (1975).
[50]   C. E. Schwartz and J. M. Smith, *Ind. Eng. Chem.* **45**, 1209 (1953).
[51]   M. Kimura, K. Nono and T. Kaneda, *Kagaku Kogaku* **19**, 397 (1955).
[52]   L. H. S. Roblee, R. M. Baird and J. W. Tierney, *AIChE J.* **4**, 460 (1958).
[53]   K. K. Pillai, *Chem. Eng. Sci.* **32**, 59 (1977).
[54]   S. Yagi, D. Kunii and N. Wakao, *AIChE J.* **6**, 543 (1960).
[55]   K. W. McHenry and R. H. Wilhelm, *AIChE J.* **3**, 83 (1957).
[56]   R. Aris and N. R. Amundson, *AIChE J.* **3**, 280 (1957).
[57]   P. V. Danckwerts, *Chem. Eng. Sci.* **2**, 1 (1953).
[58]   H. Kramers and G. Alberda, *Chem. Eng. Sci.* **2**, 173 (1953).

6 Thermal Response Measurements

THE TECHNIQUES of parameter estimation from tracer input–response signals, as discussed in Chapter 1, may be applied to the estimation of heat transfer parameters from thermal responses. The Dispersion–Concentric model (D–C model), based on the assumption that fluid is in the dispersed plug flow mode and that the intraparticle temperature/concentration profile has radial symmetry or is concentric, has been used widely for the analysis of unsteady-state heat transfer in packed beds as well as for adsorption and catalytic reaction systems.

For a bed of inert spherical particles, the unsteady-state heat balance equations based on the D–C model are

$$\frac{\partial T_F}{\partial t} = \alpha_{ax} \frac{\partial^2 T_F}{\partial x^2} - U \frac{\partial T_F}{\partial x} - \frac{a}{\epsilon_b C_F \rho_F} k_S \left(\frac{\partial T_S}{\partial r} \right)_R \tag{6.1}$$

$$\frac{\partial T_S}{\partial t} = \frac{k_S}{C_S \rho_S} \frac{1}{r^2} \frac{\partial}{\partial r} \left(r^2 \frac{\partial T_S}{\partial r} \right) \tag{6.2}$$

with

$$k_S \frac{\partial T_S}{\partial r} = h_p (T_F - T_S) \qquad \text{at } r = R \tag{6.2a}$$

where

a = particle surface area per unit volume of packed bed
C_F = specific heat of fluid
C_S = specific heat of solid particle
h_p = particle-to-fluid heat transfer coefficient
k_S = thermal conductivity of solid particle

R = particle radius, $D_p/2$
T_F = temperature of fluid
T_S = temperature of solid particle
U = interstitial fluid velocity
α_{ax} = axial fluid thermal dispersion coefficient
ϵ_b = bed void fraction
ρ_F = density of fluid
ρ_S = density of solid particle.

The rate parameters involved in unsteady-state heat transfer are usually α_{ax}, h_p and k_S. When conductivity data of solid particles are available. the parameters remaining to be determined are α_{ax} and h_p.

In the past, it has been assumed that axial fluid thermal dispersion coefficients are given by Eq. (6.3), by analogy with Eq. (2.29) for mass dispersion,

$$\begin{aligned}
\alpha'_{ax} &= (0.6 \sim 0.8)\,\alpha_F && \text{for } Re < 1 \\
&= (0.6 \sim 0.8)\,\alpha_F + 0.5 D_p U && \text{for } Re > 5
\end{aligned} \qquad (6.3)$$

where α_F is the thermal diffusivity of fluid. However, in 1974, Gunn and De Souza [1] found, from frequency response measurements, that the values of α_{ax} appeared in Eq. (6.1) were much larger than α'_{ax} predicted using Eq. (6.3).

As will be shown later, if the α'_{ax} values are assumed in the analysis of thermal response measurements, anomalously low particle-to-fluid heat transfer coefficients are obtained at low flow rates; moreover, the heat transfer coefficients, thus evaluated, tend to decrease continuously with decreasing flow rate. This implies that no heat transfer will occur between a particle and its surrounding envelope of fluid at zero or some other very low flow rate. This is against the nature of heat transfer, unless thermal equilibrium is present. On the other hand, if the values of α_{ax} obtained by Gunn and De Souza are assumed, the particle-to-fluid heat transfer coefficients, thus calculated, show no such anomalous trends at low Reynolds number. In fact, Gunn and De Souza observed that with their α_{ax} values the Nusselt numbers seemed to decrease to the asymptotic value of ten with a decrease in Reynolds number. The technique employed by Gunn and De Souza, and their findings concerning the axial thermal dispersion coefficient have led to significant developments in heat transfer in packed bed systems.

In the following sections we shall outline the techniques for parameter measurement by thermal response. The prediction of fluid thermal dispersion coefficients from axial effective thermal conductivity by model comparison will also be illustrated.

6.1   Frequency Response Measurements of Gunn and De Souza

In applying the technique of parameter measurement by thermal frequency response, Eqs. (6.1) and (6.2) based on the D-C model are solved under the following conditions:

$$T_F^I(t) = A_\omega^I \cos \omega t \qquad \text{at } x = 0 \text{ (within a bed)}$$

$$T_F = 0 \qquad \text{at } x = \infty$$

where A_ω^I is the amplitude of the input signal, $T_F^I(t)$, with frequency ω. The stationary-state solution of the response signal, $T_F^{II}(t)$, at $x = L$, is

$$T_F^{II}(t) = A_\omega^{II} \cos(\omega t - \theta L) \qquad (6.4)$$

where A_ω^{II} is the amplitude of the response signal, which, for a bed of inert particles, is defined as:

$$A_\omega^{II} = A_\omega^I \exp\left[\left(\frac{U}{2\alpha_{ax}} - \theta\eta\right) L\right] \qquad (6.4a)$$

and

$$\theta = \left\{\left(\frac{U^2}{8\alpha_{ax}^2} + \frac{p}{2\alpha_{ax} C_F \rho_F}\right) [(1 + \gamma^2)^{1/2} - 1]\right\}^{1/2} \qquad (6.4b)$$

$$\eta = \left[\frac{(1 + \gamma^2)^{1/2} + 1}{(1 + \gamma^2)^{1/2} - 1}\right]^{1/2} \qquad (6.4c)$$

$$\gamma = \frac{4\alpha_{ax}(q + \omega C_F \rho_F)}{C_F \rho_F U^2 + 4p\alpha_{ax}} \qquad (6.4d)$$

$$p = B_\omega \{ 2\phi_\omega'^2 [\cosh(2\phi_\omega') + \cos(2\phi_\omega')] - (H-1)[\cosh(2\phi_\omega') - \cos(2\phi_\omega')]$$

$$+ \phi_\omega'(H-2)[\sinh(2\phi_\omega') + \sin(2\phi_\omega')] \} \tag{6.4e}$$

$$q = HB_\omega \phi_\omega' [\sinh(2\phi_\omega') - \sin(2\phi_\omega')] \tag{6.4f}$$

$$B_\omega = \frac{\dfrac{k_S aH}{R\epsilon_b}}{\{(H-1)^2[\cosh(2\phi_\omega') - \cos(2\phi_\omega')] + 2\phi_\omega'^2[\cosh(2\phi_\omega') + \cos(2\phi_\omega')]}$$

$$+ 2\phi_\omega'(H-1)[\sinh(2\phi_\omega') - \sin(2\phi_\omega')]\} \tag{6.4g}$$

$$\phi_\omega' = R \left(\frac{\omega C_S \rho_S}{2k_S} \right)^{1/2} \tag{6.4h}$$

$$H = \frac{h_p R}{k_S}. \tag{6.4i}$$

Hence, the amplitude ratio, A_ω, of $T_F^{II}(t)$ to $T_F^{I}(t)$ is then

$$A_\omega = \frac{A_\omega^{II}}{A_\omega^{I}}$$

$$= \exp\left[\left(\frac{U}{2\alpha_{ax}} - \theta\eta \right) L \right]. \tag{6.5}$$

When $2\phi_\omega' < 1$, Eqs. (6.4e) and (6.4f) become

$$p = \omega^2 (C_S \rho_S)^2 \frac{aR}{3\epsilon_b} \left(\frac{R}{3h_p} + \frac{R^2}{15k_S} \right) \tag{6.6}$$

$$q = \omega C_S \rho_S \frac{aR}{3\epsilon_b}. \tag{6.7}$$

Gunn and De Souza [1] made thermal frequency response measurements in packed beds of glass and metal spheres over the range of Reynolds number from 0.05 to 330 for air flow. In their analysis, first, they computed theoretical amplitude ratios from Eq. (6.5) with various assumed values of α_{ax}, h_p and k_S, and then compared these with measured values. At low Reynolds numbers, say $Re < 1$, the amplitude ratio was found to be sensitive only to α_{ax}. In fact, at low flow rates and intermediate frequencies, γ^2 becomes much greater than one. Using this and Eqs. (6.6) and (6.7), Eq. (6.5) simplifies to

$$\ln A_\omega = \frac{LU}{2\alpha_{ax}} - \left(1 + \frac{aR}{3\epsilon_b} \frac{C_S\rho_S}{C_F\rho_F}\right)^{1/2} \left(\frac{\omega}{2\alpha_{ax}}\right)^{1/2} L. \tag{6.8}$$

According to Eq. (6.8), a graph of $\ln A_\omega$ versus $\omega^{1/2}$ is linear, and the value of α_{ax} may be found from the intercept or the slope of the straight line. Equation (6.8) also shows that the amplitude ratio depends only upon α_{ax}, and is independent of h_p and k_S. In fact, the imposition of a relatively slow sinusoidal change, at low flow rates, allows a particle and its surrounding envelope of fluid to reach thermal equilibrium, so that the attenuation of the temperature wave is governed by the axial thermal dispersion alone. At higher Reynolds number, thermal response is found to be equally sensitive to both α_{ax} and h_p. The sensitivity of the response to k_S is generally low, particularly for high thermal conductivity particles.

Values of α_{ax}, obtained by Gunn and De Souza [1], are plotted in terms of $D_p U/\alpha_{ax}$ in Figure 6.1. Included in the graph are the mass dispersion data reported by Edwards and Richardson [2] and Gunn and Pryce [3]. These mass dispersion data, which were obtained for argon–air systems (with a Schmidt number of 0.77) in packed beds with $\epsilon_b = 0.36$ to 0.38, are in good agreement with the solid line representing Eq. (6.3) based on $Pr = 0.77$ and $\epsilon_b = 0.37$. In any case, the α_{ax} values Gunn and De Souza determined from thermal responses are much greater than the α'_{ax} values estimated according to mass dispersion analogy, except those at high Reynolds numbers.

6.2   Parameter Estimation from One-shot Thermal Input

Parameter estimation may be made with a one-shot heat input provided that the temperature–time curves are measured at two points downstream. If the temperature signal measured at the first downstream point is $T_F^I(t)$, the response signal, $T_F^{II}(t)$, at a distance L from the first measuring point

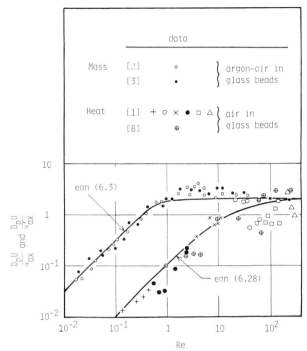

FIGURE 6.1 Comparison of $D_p U/\alpha_{ax}$ and $D_p U/\alpha'_{ax}$.

is obtained by solving Eqs. (6.1) and (6.2) under the following conditions:

$$T_F = T_S = 0 \qquad \text{at } t = 0$$

$$T_F = T_F^I(t) \qquad \text{at } x = 0 \text{ (within a bed)}$$

$$T_F = 0 \qquad \text{at } x = \infty.$$

There are two mathematical techniques for the solution of the response signal (refer to Section 1.1.6).

6.2.1 Prediction of Response Signal Using a Convolution Integral

By using a convolution integral, the response signal, $T_F^{II}{}_{calc}(t)$, is

$$T^{\mathrm{II}}_{\mathrm{F\ calc}}(t) = \int_0^t T^{\mathrm{I}}_{\mathrm{F\ expt}}(\xi)\, f(t-\xi)\, \mathrm{d}\xi \tag{6.9}$$

where $f(t)$ is the Laplace inversion of the transfer function. The transfer function, $F(s)$, of an inert bed within the distance L is

$$F(s) = \exp\left\{\frac{LU}{2\alpha_{\mathrm{ax}}}\,[1-(1+B)^{1/2}]\right\} \tag{6.10}$$

where

$$B = \frac{4\alpha_{\mathrm{ax}}}{U^2}\left[s + \left(\frac{k_S a}{\epsilon_b RC_F \rho_F}\right)\left(\cfrac{1}{\cfrac{k_S}{h_p R} + \cfrac{1}{\phi_S \coth \phi_S - 1}}\right)\right] \tag{6.10a}$$

$$\phi_S = R\left(\frac{C_S \rho_S}{k_S}\, s\right)^{1/2}. \tag{6.10b}$$

The inversion is expressed, in terms of a Fourier series, over a period of 2τ, as:

$$f(t) = \frac{1}{2\tau} + \frac{1}{\tau}\sum_{n=1}^{\infty}\left(R_n \cos\frac{n\pi t}{\tau} - I_n \sin\frac{n\pi t}{\tau}\right) \tag{6.11}$$

where 2τ is a period of time longer than the time required for the tailing portion of the measured response signal to vanish; R_n and I_n are the real and imaginary parts of $F(in\pi/\tau)$, respectively, i.e.

$$F\left(i\frac{n\pi}{\tau}\right) = R_n + iI_n \tag{6.12}$$

and

$$R_n = \left\{\exp\left[\left(\frac{U}{2\alpha_{\mathrm{ax}}} - \theta\eta\right)L\right]\cos(\theta L)\right\}_{\omega = n\pi/\tau} \tag{6.13}$$

$$I_n = -\left\{ \exp\left[\left(\frac{U}{2\alpha_{ax}} - \theta\eta\right) L\right] \sin\left(\theta L\right)\right\}_{\omega = n\pi/\tau} \tag{6.14}$$

where θ and η are defined by Eqs. (6.4b) and (6.4c). With these R_n and I_n values the response signal is then calculated from Eqs. (6.9) and (6.11):

$$T_{F\,calc}^{II}(t) = \frac{1}{2\tau} \int_0^t T_{F\,expt}^{I}(\xi)\, d\xi$$

$$+ \frac{1}{\tau} \sum_{n=1}^{\infty} \left[R_n \int_0^t T_{F\,expt}^{I}(\xi) \cos\frac{n\pi}{\tau}(t - \xi)\, d\xi \right.$$

$$\left. - I_n \int_0^t T_{F\,expt}^{I}(\xi) \sin\frac{n\pi}{\tau}(t - \xi)\, d\xi \right]. \tag{6.15}$$

6.2.2 Prediction of Response Signal by Fourier Series

The response signal, expressed by a Fourier series over the period 2τ, is

$$T_{F\,calc}^{II}(t) = \frac{a_0^\dagger}{2} + \sum_{n=1}^{\infty} \left(a_n^\dagger \cos\frac{n\pi t}{\tau} + b_n^\dagger \sin\frac{n\pi t}{\tau} \right). \tag{6.16}$$

Similar to the derivation in Section 1.1.6.2, the Fourier coefficients are determined by

$$a_n^\dagger - i b_n^\dagger = (a_n - i b_n)\, F\left(i\frac{n\pi}{\tau}\right) \tag{6.17}$$

where

$$a_n = \frac{1}{\tau} \int_0^{2\tau} T_{F\,expt}^{I} \cos\frac{n\pi t}{\tau}\, dt \tag{6.17a}$$

$$b_n = \frac{1}{\tau} \int\limits_0^{2\tau} T^I_{F \, \text{expt}} \sin \frac{n\pi t}{\tau} \, dt. \qquad (6.17b)$$

Following the method discussed in Section 1.1.6, the predicted response signal, $T^{II}_{F \, \text{calc}}(t)$, is then compared with the measured signal $T^{II}_{F \, \text{expt}}(t)$. Parameters are determined by minimizing the difference between the two response signals, for example, using the root-mean-square-error, ϵ, defined by Eq. (6.18). (The Fortran programs listed in Appendix B can easily be modified for use in the computations of the thermal response signal and the root-mean-square-error.)

$$\epsilon = \left[\frac{\displaystyle\int\limits_0^{2\tau} (T^{II}_{F \, \text{expt}} - T^{II}_{F \, \text{calc}})^2 \, dt}{\displaystyle\int\limits_0^{2\tau} (T^{II}_{F \, \text{expt}})^2 \, dt} \right]^{1/2}. \qquad (6.18)$$

6.2.3   Determination of Particle-to-fluid Heat Transfer Coefficients and Axial Fluid Thermal Dispersion Coefficients

The determination of the heat transfer parameters, particle-to-fluid heat transfer coefficient and the axial fluid thermal dispersion coefficient, is demonstrated in Example 6.1.

Example 6.1

Thermal response measurements were made using a bed of glass beads. Table 6.1 lists the experimental conditions and the measured input and response signals (data for Runs 2 and 3 are from Shen *et al.* [4]). Find the heat transfer parameters.

SOLUTION

For illustration purposes, the normalized input and response signals (similar to Eq. 1.50) for Run 2 with $Re = 17.6$, are shown in Figure 6.2. The response signals computed with various assumed values of the three

TABLE 6.1
Experimental conditions and temperature signals recorded in a bed of glass beads.

Glass beads: $D_p = 0.5$, 1.3 and 2.7 mm; $C_S = 670$ J kg^{-1} K^{-1}; $\rho_S = 2500$ kg m^{-3}; $k_S = 0.88$ W m^{-1} K^{-1}.
Fluid: air at 0.1 MPa.
Average temperature of the air in the bed = 20°C; $k_F = 0.026$ W m^{-1} K^{-1}.
The response signals were measured in the packed beds at 4 cm from the bed exit.
Input and response signals measured (before normalized; the figures are ten times the actual temperature increase (K) above the room temperature).

Time	Run 1		Run 2		Run 3	
	$D_p = 0.5$ mm $\epsilon_b = 0.39$ $L = 1.1$ cm $Re = 0.54$		$D_p = 1.3$ mm $\epsilon_b = 0.39$ $L = 1.3$ cm $Re = 17.6$		$D_p = 2.7$ mm $\epsilon_b = 0.40$ $L = 2.4$ cm $Re = 229$	
$n\Delta t$ (s)						
n	T_F^I	T_F^{II}	T_F^I	T_F^{II}	T_F^I	T_F^{II}
0	0.0	—	0.0	—	0.0	0.0
1	12.5	—	0.9	—	26.7	0.7
2	33.0	0.0	0.9	—	67.2	3.0
3	52.5	3.3	0.9	—	95.0	4.6
4	60.0	9.5	1.0	—	122.0	8.0
5	63.2	15.2	1.8	—	129.5	12.0
6	65.2	21.0	3.1	—	127.9	16.4
7	64.9	26.9	5.9	—	122.0	20.8
8	64.0	30.1	9.8	—	112.9	25.3
9	62.2	33.2	13.9	—	103.9	29.9
10	60.6	36.6	19.8	—	94.6	34.1
11	58.0	39.0	25.9	—	85.4	38.1
12	55.4	41.4	32.9	0.0	76.9	41.9
13	52.8	42.4	39.8	0.1	68.9	45.3
14	50.4	42.9	46.3	0.2	61.4	47.9
15	47.8	43.0	52.8	0.5	54.4	50.2
16	45.4	43.0	58.8	0.8	48.7	52.4
17	43.2	42.9	64.3	1.0	42.8	53.6
18	41.0	42.8	68.7	1.3	37.9	54.6
19	38.8	42.1	72.5	1.9	32.9	55.2
20	36.9	41.0	75.0	2.8	29.3	55.6
21	35.0	40.1	77.1	3.8	25.9	55.6
22	33.1	39.0	78.0	4.4	22.4	55.6
23	31.5	38.3	78.6	5.8	19.8	54.8
24	29.8	37.3	78.6	6.9	17.2	53.8
25	27.7	36.2	77.7	8.7	15.1	52.6
26	25.9	35.1	76.5	10.1	13.3	51.3
27	24.9	33.7	74.7	12.1	11.6	49.1
28	23.9	32.7	72.7	14.3	10.2	47.5
29	22.9	31.6	70.2	16.7	9.0	45.6

TABLE 6.1 (Continued)

Time	Run 1 $D_p = 0.5$ mm $\epsilon_b = 0.39$ $L = 1.1$ cm $Re = 0.54$		Run 2 $D_p = 1.3$ mm $\epsilon_b = 0.39$ $L = 1.3$ cm $Re = 17.6$		Run 3 $D_p = 2.7$ mm $\epsilon_b = 0.40$ $L = 2.4$ cm $Re = 229$	
$n\Delta t$ (s) n	T_F^I	T_F^{II}	T_F^I	T_F^{II}	T_F^I	T_F^{II}
30	21.6	30.5	67.7	18.9	7.9	43.6
31	20.4	29.3	64.9	20.9	6.9	41.7
32	19.5	28.2	62.1	22.9	5.9	39.6
33	18.4	27.0	59.3	25.3	5.3	37.7
34	17.5	25.9	56.1	27.7	4.8	35.6
35	16.2	24.8	53.4	29.8	4.3	33.5
36	15.3	23.7	50.5	31.9	3.7	31.5
37	14.5	22.7	47.6	34.3	3.2	29.5
38	13.7	21.6	44.8	36.3	2.7	27.5
39	13.1	20.7	42.5	37.9	2.5	25.7
40	12.6	20.2	39.6	39.9	2.2	24.3
41	12.1	19.6	37.4	41.8	2.0	22.5
42	11.6	18.8	34.5	42.8	1.7	20.7
43	10.8	18.1	32.5	44.3	1.6	19.3
44	10.3	17.5	30.4	45.5	1.4	17.8
45	9.6	16.6	28.5	46.5	1.2	16.6
46	9.3	16.0	26.5	47.4	1.1	15.5
47	9.0	15.4	24.5	47.7	1.0	14.4
48	8.7	14.6	22.9	48.1	0.9	13.3
49	8.4	14.2	21.5	48.7	0.8	12.2
50	8.0	13.6	19.8	48.7	0.7	11.3
51	7.7	12.9	18.4	48.7	0.7	10.4
52	7.4	12.4	17.3	48.7	0.7	9.5
53	7.0	12.0	16.2	48.7	0.7	8.6
54	6.7	11.5	15.0	48.3	0.6	7.7
55	6.3	11.1	13.8	47.7	0.6	7.1
56	6.0	10.6	12.6	47.1	0.6	6.5
57	5.7	10.2	11.5	46.4	0.5	6.0
58	5.4	9.8	10.4	45.1	0.5	5.5
59	5.2	9.3	9.9	44.3	0.5	5.0
60	5.0	8.9	9.3	43.4	0.5	4.5
61	4.7	8.5	8.8	42.4	0.4	4.3
62	4.5	8.1	8.2	41.3	0.4	3.9
63	4.4	7.8	7.5	40.3	0.3	3.5
64	4.3	7.5	6.8	39.1	0.3	3.2
65	4.0	7.2	6.3	37.9	0.3	3.0
66	3.8	7.0	5.8	36.7	0.2	2.7
67	3.7	6.7	5.5	35.5	0.2	2.4
68	3.5	6.4	5.2	34.3	0.1	2.3
69	3.3	6.2	4.8	33.1	0.1	2.1

TABLE 6.1 (Continued)

Time	Run 1		Run 2		Run 3	
	$D_p = 0.5$ mm $\epsilon_b = 0.39$ $L = 1.1$ cm $Re = 0.54$		$D_p = 1.3$ mm $\epsilon_b = 0.39$ $L = 1.3$ cm $Re = 17.6$		$D_p = 2.7$ mm $\epsilon_b = 0.40$ $L = 2.4$ cm $Re = 229$	
$n\Delta t$ (s)						
n	T_F^I	T_F^{II}	T_F^I	T_F^{II}	T_F^I	T_F^{II}
70	3.2	6.0	4.4	31.9	0.1	2.0
71	2.9	5.7	4.2	30.6	0.0	1.9
72	2.7	5.5	4.0	29.4	–	1.8
73	2.5	5.2	3.7	28.2	–	1.7
74	2.4	5.0	3.4	27.0	–	1.5
75	2.1	4.7	3.2	25.8	–	1.4
76	2.0	4.4	2.9	24.6	–	1.3
77	1.9	4.1	2.7	23.4	–	1.2
78	1.8	3.9	2.5	22.3	–	1.1
79	1.7	3.7	2.3	21.3	–	1.1
80	1.5	3.5	2.1	20.3	–	1.0
81	1.4	3.3	2.0	19.3	–	0.9
82	1.2	3.2	1.8	18.3	–	0.9
83	1.2	3.1	1.6	17.4	–	0.8
84	1.1	3.0	1.5	16.4	–	0.8
85	0.9	2.9	1.4	15.5	–	0.7
86	0.9	2.7	1.3	14.6	–	0.7
87	0.9	2.6	1.1	13.9	–	0.6
88	0.8	2.5	1.0	13.2	–	0.5
89	0.7	2.4	0.9	12.6	–	0.5
90	0.6	2.2	0.8	11.9	–	0.4
91	0.6	2.0	0.7	11.3	–	0.4
92	0.5	1.9	0.6	10.6	–	0.3
93	0.4	1.8	0.5	10.0	–	0.2
94	0.4	1.6	0.4	9.4	–	0.2
95	0.2	1.5	0.4	8.9	–	0.1
96	0.1	1.4	0.3	8.4	–	0.1
97	0.1	1.3	0.3	7.8	–	0.0
98	0.1	1.2	0.3	7.3	–	–
99	0.0	1.1	0.3	6.8	–	–
100	–	1.0	0.2	6.5	–	–
101	–	0.9	0.2	6.2	–	–
102	–	0.8	0.2	5.9	–	–
103	–	0.6	0.1	5.6	–	–
104	–	0.4	0.1	5.2	–	–
105	–	0.3	0.1	4.9	–	–
106	–	0.2	0.1	4.6	–	–
107	–	0.0	0.1	4.3	–	–
108	–	–	0.0	4.1	–	–
109	–	–	–	3.9	–	–

TABLE 6.1 (Continued)

Time	Run 1 $D_p = 0.5$ mm $\epsilon_b = 0.39$ $L = 1.1$ cm $Re = 0.54$		Run 2 $D_p = 1.3$ mm $\epsilon_b = 0.39$ $L = 1.3$ cm $Re = 17.6$		Run 3 $D_p = 2.7$ mm $\epsilon_b = 0.40$ $L = 2.4$ cm $Re = 229$	
$n\Delta t$ (s)						
n	T_F^I	T_F^{II}	T_F^I	T_F^{II}	T_F^I	T_F^{II}
110	—	—	—	3.7	—	—
111	—	—	—	3.5	—	—
112	—	—	—	3.4	—	—
113	—	—	—	3.2	—	—
114	—	—	—	3.0	—	—
115	—	—	—	2.8	—	—
116	—	—	—	2.6	—	—
117	—	—	—	2.4	—	—
118	—	—	—	2.3	—	—
119	—	—	—	2.1	—	—
120	—	—	—	2.0	—	—
121	—	—	—	1.9	—	—
122	—	—	—	1.8	—	—
123	—	—	—	1.7	—	—
124	—	—	—	1.6	—	—
125	—	—	—	1.6	—	—
126	—	—	—	1.5	—	—
127	—	—	—	1.4	—	—
128	—	—	—	1.3	—	—
129	—	—	—	1.2	—	—
130	—	—	—	1.1	—	—
131	—	—	—	1.0	—	—
132	—	—	—	1.0	—	—
133	—	—	—	0.9	—	—
134	—	—	—	0.8	—	—
135	—	—	—	0.7	—	—
136	—	—	—	0.6	—	—
137	—	—	—	0.5	—	—
138	—	—	—	0.4	—	—
139	—	—	—	0.3	—	—
140	—	—	—	0.2	—	—
141	—	—	—	0.1	—	—
142	—	—	—	0.0	—	—
Time interval Δt (s)	60	60	2	2	1	1

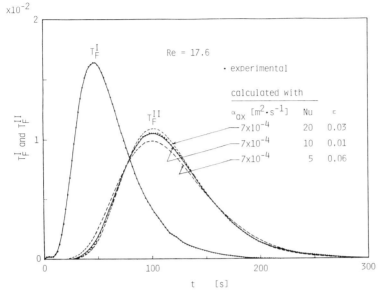

FIGURE 6.2   Normalized input and response signals measured and response signals predicted, for Run 2 in Example 6.1.

parameters, α_{ax}, h_p (in terms of $Nu = h_p D_p / k_F$) and k_S, are compared with the measured response signal. The difference between the two curves is then evaluated using Eq. (6.18).

Figures 6.3(a)-(c) show the relationships between the three parameters, α_{ax}, Nu and k_S at $Re = 0.54$, 17.6 and 229, respectively. As shown, particle thermal conductivity has little effect on the thermal responses, when the conductivity is high and/or the flow rate is low.

In this example, the values of k_S, C_S and ρ_S for the glass beads are provided. The unknown parameters are, therefore, α_{ax} and h_p. At the average bed temperature of 20°C, the thermal conductivity, k_F, of air is 0.026 W m^{-1} K^{-1}. The thermal conductivity of the glass particles is $k_S = 0.88$ W m^{-1} K^{-1} or $k_S/k_F = 34$; therefore, according to Figure 5.9, $k_e^o/k_F = 6.4$ for the glass-air system. The values of α_{ax} estimated from

Eq. (6.28) are compared with α'_{ax} predicted from Eq. (6.3) below:

Re	α_{ax} (m² s⁻¹)	α'_{ax} (m² s⁻¹)
0.54	3.5×10^{-4}	$(0.23 \sim 0.27) \times 10^{-4}$
17.6	6.7×10^{-4}	$(3.4 \sim 3.5) \times 10^{-4}$
229	47×10^{-4}	43×10^{-4}

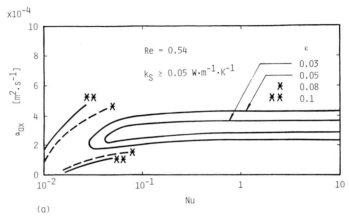

(a)

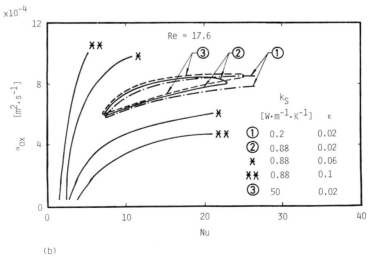

(b)

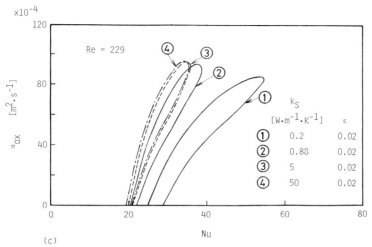

(c)

FIGURE 6.3   Relationships between α_{ax}, Nu and k_S, for glass beads, for Example 6.1: (a) $Re = 0.54$; (b) $Re = 17.6$; (c) $Re = 229$.

From the contours with the least root-mean-square-errors (or valley in the three-dimensional error map) drawn in Figures 6.3(a)-(c), the particle-to-gas heat transfer coefficients can be estimated using the above α_{ax} values as:

Re	Nu	ϵ
0.54	$0.1 \sim \infty$	0.03
17.6	$8 \sim 12$	0.02
229	$26 \sim 30$	0.02

At $Re = 0.54$, any Nusselt number, in the range 0.1 to ∞, gives good agreement between the predicted response curves and the measured signal. According to Eq. (8.20), however, the Nusselt number at this Reynolds number is expected to be 2.7. The Nusselt numbers determined at $Re = 17.6$ and 229 concur well with those predicted from Eq. (8.20), $Nu = 8$ at $Re = 17.6$ and $Nu = 28$ at $Re = 229$. For Run 2 with $Re = 17.6$, the effect of Nu on the shape of the predicted response curves is illustrated in Figure 6.2. It can be clearly seen that the response signal, computed with $Nu = 10$, matches the measured signal fairly well; while the curves predicted, with $Nu = 20$ or 5, differ considerably from the measured signal.

Figure 6.3(a) reveals that, at $Re = 0.54$, α_{ax} is almost independent of Nu if Nu is greater than about 0.1. Within a confidence range of $\epsilon = 0.03$, the value of α_{ax} falls in the range $(2.8 \sim 3.6) \times 10^{-4}$ m^2 s^{-1}, which agrees well with the value of 3.5×10^{-4} m^2 s^{-1} estimated from Eq. (6.28), but differs significantly from the α'_{ax} value, $(0.23 \sim 0.27) \times 10^{-4}$ m^2 s^{-1}, predicted using Eq. (6.3).

(End of Example)

As illustrated in the above example, definite values of the Nusselt number can not be determined from thermal response measurements at low flow rates. This arises from the fact that, when the flow rate is low, thermal equilibrium is nearly attained between the particle and its surrounding envelope of fluid, such that particle-to-fluid heat transfer makes little contribution to the overall heat transfer.

Figures 6.3(a) and (b) show that if the α'_{ax} values from Eq. (6.3) are assumed, the root-mean-square-error, ϵ, is large, i.e. the agreement between the measured and predicted response curves is not good. In any case if the α'_{ax} values are assumed, the Nusselt numbers estimated from the contours labeled * and ** are much smaller than those obtained using the large values of α_{ax} from Eq. (6.28). At high flow rates, as seen from Figure 6.3(c), the Nusselt number is little affected by the axial fluid thermal dispersion coefficient. Also, the α_{ax} value predicted from Eq. (6.28) is close to the α'_{ax} value estimated from Eq. (6.3).

6.2.4   Determination of Particle Thermal Conductivities

If particle-to-fluid heat transfer coefficients are known, the particle thermal conductivity can be determined from thermal response measurements.

The difference in the second central moment or variance, σ^2_H, between the response and input signals is

$$\sigma^2_H = \frac{2L}{U} \left[\gamma_1 + \alpha_{ax} \frac{(1 + \gamma_0)^2}{U^2} \right] \qquad (6.19)$$

where

$$\gamma_0 = \frac{aR}{3\epsilon_b} \frac{C_S \rho_S}{C_F \rho_F} \qquad (6.19a)$$

$$\gamma_1 = \frac{aR}{3\epsilon_b} \frac{(C_S\rho_S)^2}{C_F\rho_F} \left(\frac{R^2}{15k_S} + \frac{R}{3h_p} \right).$$ (6.19b)

As discussed in Section 1.2.2, the least-error line in the middle of the contour corresponds to σ_H^2 = constant or $d(\sigma_H^2) = 0$. The slope of the line in a graph of α_{ax} versus k_S is then

$$\frac{d\alpha_{ax}}{dk_S} = \frac{1}{20\epsilon_b aRC_F\rho_F} \left[\frac{\gamma_0}{(1+\gamma_0)} \frac{k_F}{k_S} (Pr)(Re) \right]^2.$$ (6.20)

According to Eq. (6.20), the least-error line is nearly horizontal, when the flow rate is sufficiently low and/or when the solid thermal conductivity is high. With an increase in flow rate and/or a reduction in solid thermal conductivity, the slope of the least-error line is increased and eventually the line becomes vertical above a certain flow rate for low thermal conductivity solids. The particle thermal conductivity is determined, therefore from the bottom of the overlapping valleys of the contour maps generated from response measurements made at both high and low flow rates. If the particle thermal conductivity is expected to be high, then, response measurement at extremely high flow rates is required in order to obtain a steep valley in an error map. The following example illustrates the determination of particle thermal conductivity from one-shot thermal input.

Example 6.2

Thermal response measurements were carried out by Shen *et al.* [5] with a bed of spherical polystyrene foam particles. Table 6.2 lists the experimental conditions and the input and response signals measured. Find the thermal conductivity of polystyrene foam particles.

SOLUTION

Equation (6.28) may be rewritten as:

$$\delta_H = \frac{\alpha_{ax} - \alpha_{ax}^o}{D_p U}$$ (6.21)

TABLE 6.2

Experimental conditions and temperature signals recorded in a bed of polystyrene foam particles.

Spherical polystyrene foam: $D_p = 3.0$ mm; $C_S = 1260$ J kg^{-1} K^{-1}; $\rho_S = 60$ kg m^{-3}.
Bed void fraction: $\epsilon_b = 0.42$.
Distance between the input and response signal measuring points: $L = 3.0$ cm.
Fluid: air at 0.1 MPa.
Average temperature of the air in the bed $= 20°C$; $k_F = 0.026$ W m^{-1} K^{-1}.
Input and response measured (before normalized; the figures are ten times the actual temperature increase (K) above the room temperature).

Time $n\Delta t$ (s) n	Run 1 $Re = 11$ T_F^I	T_F^{II}	Run 2 $Re = 26$ T_F^I	T_F^{II}	Run 3 $Re = 107$ T_F^I	T_F^{II}	Run 4 $Re = 416$ T_F^I	T_F^{II}
0	0.0	–	0.0	–	0.0	–	0.0	–
1	0.2	–	4.0	0.0	1.2	0.0	0.3	0.0
2	1.5	–	20.0	0.4	6.8	0.1	6.0	0.7
3	4.2	–	49.0	3.0	18.0	2.9	19.0	4.8
4	8.6	–	79.0	9.1	35.0	7.2	37.9	14.0
5	15.2	0.0	103.0	21.0	56.2	15.4	60.0	28.0
6	23.4	0.3	117.1	38.0	78.3	25.6	85.0	45.5
7	32.3	0.7	120.8	57.0	98.0	38.5	108.0	64.0
8	41.3	1.6	115.0	73.0	114.4	53.8	130.1	83.2
9	49.2	3.2	104.0	87.9	124.7	69.5	150.0	103.5
10	55.9	5.4	89.0	98.1	129.2	82.0	166.6	122.1
11	61.2	8.4	75.0	100.7	130.1	96.0	182.8	140.3
12	63.7	12.3	61.9	99.8	127.1	104.3	193.0	157.6
13	64.8	16.5	50.9	93.7	120.5	111.2	195.0	170.5
14	64.9	21.7	41.8	83.9	112.4	115.1	190.0	178.0
15	63.5	26.8	34.0	73.9	102.8	116.2	180.0	180.2
16	61.1	32.1	29.0	63.5	92.2	115.4	166.1	177.1
17	57.8	37.4	24.5	53.7	83.8	111.9	151.4	170.4
18	54.1	42.1	20.8	44.9	75.0	106.4	138.2	162.0
19	50.4	46.2	17.6	37.9	66.7	100.1	123.0	151.8
20	46.9	49.2	15.0	31.1	58.4	92.5	108.0	140.0
21	43.1	51.7	12.6	25.8	52.0	85.0	95.0	127.8
22	39.7	53.2	11.0	20.9	45.9	77.7	83.9	116.2
23	36.6	54.2	9.8	18.2	40.3	70.5	73.6	104.7
24	33.4	54.3	8.7	15.7	35.6	63.3	64.7	94.0
25	30.7	54.5	7.9	13.8	31.2	56.1	57.0	84.1
26	27.9	53.4	7.0	11.7	28.3	50.4	50.1	74.6
27	25.8	51.8	6.1	10.4	25.0	44.9	44.0	66.2
28	23.8	49.3	5.8	9.0	22.4	39.3	38.8	58.6
29	21.8	47.0	5.2	7.9	20.0	34.9	33.6	51.6

TABLE 6.2 (Continued)

Time $n\Delta t$ (s) n	Run 1 $Re = 11$		Run 2 $Re = 26$		Run 3 $Re = 107$		Run 4 $Re = 416$	
	T_F^I	T_F^{II}	T_F^I	T_F^{II}	T_F^I	T_F^{II}	T_F^I	T_F^{II}
30	20.3	44.5	4.9	7.2	18.0	30.9	29.9	45.3
31	18.8	42.0	4.4	6.3	16.0	27.1	26.0	39.9
32	17.6	39.2	4.1	5.8	14.3	23.9	23.0	35.1
33	16.4	36.6	3.9	5.5	13.0	21.4	20.7	31.0
34	15.1	34.2	3.6	4.9	11.8	19.2	18.2	27.3
35	14.0	31.4	3.4	4.6	10.8	17.0	16.2	24.2
36	13.1	29.4	3.2	4.3	9.9	15.5	14.3	21.4
37	12.4	27.2	3.0	3.9	9.0	13.8	13.1	18.9
38	11.8	25.3	2.7	3.7	8.2	12.4	11.8	16.7
39	10.8	23.7	2.5	3.6	7.5	11.4	10.6	14.8
40	10.2	21.8	2.2	3.1	7.0	10.4	9.7	13.2
41	9.8	20.3	2.0	2.9	6.5	9.5	8.8	11.7
42	9.0	19.2	2.0	2.8	6.0	8.8	7.9	10.5
43	8.6	17.8	2.0	2.6	5.6	8.0	7.2	9.4
44	8.2	16.6	1.9	2.4	5.1	7.4	6.9	8.4
45	7.7	15.4	1.8	2.2	4.9	7.0	6.1	7.5
46	7.3	14.5	1.7	2.1	4.7	6.3	5.6	6.9
47	6.9	13.6	1.6	1.9	4.5	5.8	5.1	6.3
48	6.5	12.9	1.5	1.8	4.2	5.5	4.8	5.7
49	6.1	12.2	1.5	1.8	4.0	5.2	4.3	5.1
50	5.7	11.4	1.4	1.7	3.9	4.8	4.1	4.8
51	5.5	10.6	1.3	1.7	3.7	4.6	3.9	4.6
52	5.2	10.0	1.2	1.6	3.6	4.4	3.7	4.3
53	4.9	9.4	1.1	1.6	3.4	4.2	3.5	3.9
54	4.7	8.8	1.0	1.5	3.3	4.0	3.2	3.6
55	4.5	8.6	1.0	1.5	3.1	3.8	3.1	3.4
56	4.3	8.0	1.0	1.4	3.0	3.6	2.9	3.3
57	4.1	7.7	1.0	1.3	2.9	3.5	2.8	3.1
58	3.9	7.4	1.0	1.2	2.7	3.3	2.6	2.9
59	3.7	6.9	0.9	1.2	2.6	3.1	2.5	2.8
60	3.5	6.7	0.9	1.1	2.5	3.0	2.4	2.7
61	3.4	6.4	0.9	1.1	2.5	2.8	2.3	2.5
62	3.3	5.9	0.8	1.0	2.4	2.7	2.2	2.4
63	3.1	5.6	0.8	1.0	2.3	2.6	2.2	2.3
64	3.0	5.4	0.8	1.0	2.3	2.6	2.1	2.2
65	2.9	5.2	0.7	0.9	2.2	2.5	2.0	2.1
66	2.8	4.9	0.7	0.9	2.2	2.4	1.9	2.1
67	2.6	4.8	0.7	0.9	2.1	2.3	1.8	2.1
68	2.5	4.4	0.6	0.8	2.1	2.2	1.7	2.0
69	2.4	4.3	0.6	0.8	2.0	2.2	1.6	2.0

TABLE 6.2 (Continued)

Time $n\Delta t$ (s)	Run 1 $Re = 11$		Run 2 $Re = 26$		Run 3 $Re = 107$		Run 4 $Re = 416$	
n	T_F^I	T_F^{II}	T_F^I	T_F^{II}	T_F^I	T_F^{II}	T_F^I	T_F^{II}
70	2.4	4.1	0.6	0.8	1.9	2.1	1.5	2.0
71	2.3	3.9	0.5	0.7	1.9	2.1	1.5	1.9
72	2.2	3.7	0.5	0.7	1.8	2.0	1.4	1.9
73	2.1	3.6	0.5	0.7	1.8	2.0	1.4	1.9
74	2.0	3.4	0.4	0.6	1.7	1.9	1.4	1.9
75	1.9	3.3	0.4	0.6	1.7	1.8	1.3	1.8
76	1.8	3.2	0.4	0.5	1.6	1.8	1.3	1.8
77	1.7	3.1	0.3	0.5	1.6	1.8	1.3	1.8
78	1.6	3.0	0.3	0.5	1.5	1.7	1.3	1.7
79	1.6	2.9	0.3	0.4	1.5	1.7	1.2	1.7
80	1.5	2.8	0.2	0.4	1.4	1.6	1.2	1.7
81	1.4	2.7	0.2	0.4	1.4	1.6	1.2	1.7
82	1.4	2.5	0.2	0.3	1.3	1.5	1.1	1.6
83	1.3	2.5	0.1	0.3	1.3	1.5	1.1	1.6
84	1.3	2.4	0.1	0.3	1.3	1.4	1.1	1.6
85	1.2	2.3	0.1	0.2	1.2	1.4	1.0	1.5
86	1.2	2.1	0.0	0.2	1.1	1.3	1.0	1.5
87	1.1	2.0	—	0.2	1.1	1.3	1.0	1.5
88	1.1	1.9	—	0.1	1.0	1.2	1.0	1.5
89	1.0	1.8	—	0.1	1.0	1.2	0.9	1.4
90	0.9	1.7	—	0.1	1.0	1.1	0.9	1.4
91	0.9	1.6	—	0.0	0.9	1.1	0.9	1.4
92	0.8	1.5	—	—	0.9	1.0	0.8	1.3
93	0.8	1.4	—	—	0.9	1.0	0.8	1.3
94	0.7	1.3	—	—	0.9	0.9	0.8	1.3
95	0.7	1.2	—	—	0.8	0.9	0.7	1.2
96	0.6	1.1	—	—	0.8	0.8	0.7	1.2
97	0.6	1.0	—	—	0.8	0.8	0.7	1.1
98	0.5	0.9	—	—	0.8	0.7	0.7	1.1
99	0.4	0.8	—	—	0.7	0.7	0.6	1.1
100	0.4	0.7	—	—	0.7	0.7	0.6	1.0
101	0.3	0.6	—	—	0.7	0.6	0.6	1.0
102	0.3	0.6	—	—	0.7	0.6	0.5	1.0
103	0.2	0.5	—	—	0.6	0.6	0.5	0.9
104	0.2	0.5	—	—	0.6	0.6	0.5	0.9
105	0.1	0.4	—	—	0.6	0.6	0.4	0.9
106	0.0	0.3	—	—	0.6	0.6	0.4	0.8
107	—	0.2	—	—	0.5	0.6	0.4	0.8
108	—	0.2	—	—	0.5	0.5	0.4	0.8
109	—	0.1	—	—	0.5	0.5	0.3	0.7

TABLE 6.2 (Continued)

Time $n\Delta t$ (s) n	Run 1 $Re = 11$ T_F^I	T_F^{II}	Run 2 $Re = 26$ T_F^I	T_F^{II}	Run 3 $Re = 107$ T_F^I	T_F^{II}	Run 4 $Re = 416$ T_F^I	T_F^{II}
110	—	0.1	—	—	0.5	0.5	0.3	0.7
111	—	0.0	—	—	0.4	0.5	0.3	0.7
112	—	—	—	—	0.4	0.5	0.2	0.6
113	—	—	—	—	0.4	0.5	0.2	0.6
114	—	—	—	—	0.4	0.5	0.2	0.5
115	—	—	—	—	0.3	0.5	0.1	0.5
116	—	—	—	—	0.3	0.4	0.1	0.5
117	—	—	—	—	0.3	0.4	0.1	0.4
118	—	—	—	—	0.3	0.4	0.1	0.4
119	—	—	—	—	0.2	0.4	0.0	0.4
120	—	—	—	—	0.2	0.4	—	0.3
121	—	—	—	—	0.2	0.4	—	0.3
122	—	—	—	—	0.2	0.4	—	0.3
123	—	—	—	—	0.1	0.3	—	0.2
124	—	—	—	—	0.1	0.3	—	0.2
125	—	—	—	—	0.1	0.3	—	0.2
126	—	—	—	—	0.1	0.3	—	0.1
127	—	—	—	—	0.0	0.3	—	0.1
128	—	—	—	—	—	0.3	—	0.0
129	—	—	—	—	—	0.3	—	—
130	—	—	—	—	—	0.3	—	—
131	—	—	—	—	—	0.2	—	—
132	—	—	—	—	—	0.2	—	—
133	—	—	—	—	—	0.2	—	—
134	—	—	—	—	—	0.2	—	—
135	—	—	—	—	—	0.2	—	—
136	—	—	—	—	—	0.2	—	—
137	—	—	—	—	—	0.2	—	—
138	—	—	—	—	—	0.1	—	—
139	—	—	—	—	—	0.1	—	—
140	—	—	—	—	—	0.1	—	—
141	—	—	—	—	—	0.1	—	—
142	—	—	—	—	—	0.1	—	—
143	—	—	—	—	—	0.1	—	—
144	—	—	—	—	—	0.1	—	—
145	—	—	—	—	—	0.1	—	—
146	—	—	—	—	—	0.0	—	—
Time interval Δt (s)	2.0	2.0	2.0	2.0	0.5	0.5	0.2	0.2

where

$$\alpha_{\text{ax}}^{\text{o}} = \frac{k_{\text{e}}^{\text{o}}}{\epsilon_{\text{b}} C_{\text{F}} \rho_{\text{F}}}. \qquad (6.21\text{a})$$

The particle-to-gas heat transfer coefficients, h_{p}, are estimated from Eq. (8.20). Response signals are then predicted by assuming various values of α_{ax} and k_{S}. The predicted signals are compared with the measured signal, by evaluating the error, ϵ of Eq. (6.18), between the two signals.

Figure 6.4(a) is an error map of a plot of δ_{H} versus k_{S}. The values of δ_{H} and k_{S} are then determined from the area where all four valleys overlap. This is clearly depicted in Figure 6.4(b), which is an arithmetic mean of the four error maps. From the least-error point, labeled +, the value of δ_{H} is found to be 0.55 and k_{S} to be 0.048 W m^{-1} K^{-1}.

Similar calculations are made with Nusselt numbers 0.5 and 1.5 times the values estimated from Eq. (8.20). The average error maps shown in Figure 6.4(c) indicates that, in the response measurements with the extremely low thermal conductivity particles at high flow rates, the particle-to-gas heat transfer is a significant parameter, having a large

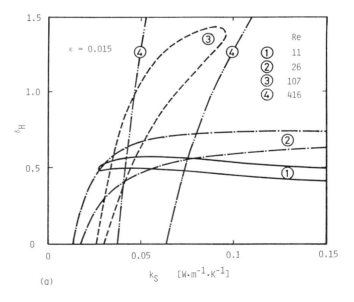

(a)

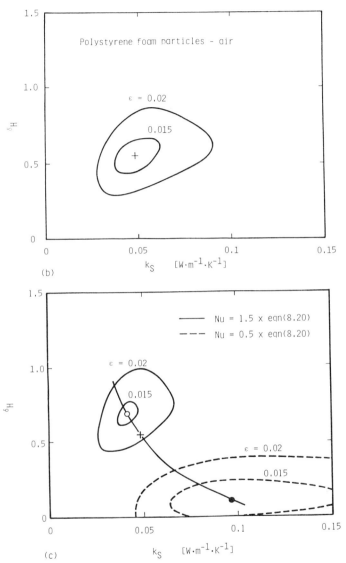

FIGURE 6.4   Error maps in the plot of δ_H versus k_S, for polystyrene foam particles, for Example 6.2: (a) effect of Re on δ_H–k_S relationship; (b) average error map for four runs; (c) effect of Nu on δ_H–k_S relationship: least-error point labeled + resulting from h_p of Eq. (8.20); • from h_p of 0.5 times Eq. (8.20); ○ from h_p of 1.5 times Eq. (8.20).

influence on the values of δ_H and k_S. The least-error point moves up and left along the locus with an increase in the Nusselt value, and down and right with decreasing Nusselt number. Equation (8.20) is the empirical correlation obtained from numerous experimental data reported in the literature. Suppose the data scattering is roughly ±50%. A 50% increase in Nusselt value, or 1.5 times the Nusselt values of Eq. (8.20), gives, as seen in Figure 6.4(c): $\delta_H = 0.69$ and $k_S = 0.042$ W m^{-1} K^{-1}. A 50% decrease in Nusselt values results in a greater effect, i.e. $\delta_H = 0.11$ and $k_S = 0.095$ W m^{-1} K^{-1}.

However, the value of δ_H can be neither as small as 0.11 nor as large as 0.69. Most previous work in the literature indicates that $\delta_H = 0.5$ (or $\delta_H = 0.45$ to 0.55), so that the thermal conductivity of the polystyrene foam particles is found, from the measurements, to be $k_S = 0.05$ W m^{-1} K^{-1} (or $k_S = 0.054$ to 0.048 W m^{-1} K^{-1}, corresponding to $\delta_H = 0.45$ to 0.55).

(End of Example)

6.3   Fluid Thermal Dispersion Coefficients

Gunn and De Souza [1] pointed out that the contribution of solid phase conduction to thermal dispersion causes the axial fluid dispersion coefficient for heat to be larger than that for mass. Quantitative interpretations of the fluid thermal dispersion and solid heat conduction relationship were subsequently furnished by Vortmeyer [6] and Wakao [7]. Vortmeyer based his analysis on a comparison of the Continuous–Solid Phase model (C–S model, see Discussion in Section 7.4) and the Single Phase model (the heterogeneous packed bed is assumed to be a homogeneous single phase). He obtained an expression which relates the axial fluid thermal dispersion coefficient to the effective solid phase conductivity. On the other hand, from a comparison of the D–C model and the Single Phase model, Wakao derived a formula relating the two heat transfer parameters. Since the C–S model, employed by Vortmeyer, is considered to be inadequate for describing the phenomenon of heat transfer in packed beds, only Wakao's interpretation will be discussed in this section.

Consider a packed bed with constant heat generation in the particles. Based on the D–C model, the steady-state heat balance equations are given as:

$$\alpha_{ax} \frac{d^2 T_F}{dx^2} - U \frac{dT_F}{dx} - \frac{a}{\epsilon_b C_F \rho_F} k_S \left(\frac{\partial T_S}{\partial r} \right)_R = 0 \qquad (6.22)$$

and

$$k_S \frac{1}{r^2} \frac{\partial}{\partial r} \left(r^2 \frac{\partial T_S}{\partial r} \right) + q_v = 0 \qquad (6.23)$$

where q_v is the rate of heat generation per unit volume of solid particle.
Solving Eq. (6.23) and substituting this into Eq. (6.22), it follows that:

$$\alpha_{ax} \frac{d^2 T_F}{dx^2} - U \frac{d T_F}{dx} + \frac{aR}{3\epsilon_b C_F \rho_F} q_v = 0. \qquad (6.24)$$

Based on the Single Phase model, on the other hand, the heat balance equation of the packed bed may be simply described in terms of the axial effective thermal conductivity, k_{eax}, as:

$$\frac{k_{eax}}{\epsilon_b C_F \rho_F} \frac{d^2 T}{dx^2} - U \frac{dT}{dx} + \frac{aR}{3\epsilon_b C_F \rho_F} q_v = 0. \qquad (6.25)$$

Under steady-state conditions, the bed temperature T is equal to the fluid temperature T_F so that a comparison of Eqs. (6.24) and (6.25) yields

$$\alpha_{ax} = \frac{k_{eax}}{\epsilon_b C_F \rho_F} \qquad (6.26)$$

or

$$\frac{\alpha_{ax}}{\alpha_F} = \frac{k_{eax}}{\epsilon_b k_F}. \qquad (6.27)$$

The effective axial thermal conductivity, discussed in Section 5.5, is given by Eq. (5.69). Substitution of Eq. (5.69) into Eq. (6.27) gives

$$\frac{\alpha_{ax}}{\alpha_F} = \frac{1}{\epsilon_b} \left[\frac{k_e^o}{k_F} + 0.5(Pr)(Re) \right]. \qquad (6.28)$$

Equation (6.28), obtained by Wakao [7], is widely employed for estimating the axial thermal dispersion coefficient, α_{ax}, in packed bed heat transfer. Values of the dispersion coefficient calculated from Eqs.

(6.28) and (6.3) agree reasonably well at high Reynolds numbers, but are significantly different from each other when the flow rates are low. In general, α_{ax} values predicted using Eq. (6.28) agree fairly well with reported experimental data. As an illustration, the experimental α_{ax} data obtained by Gunn and De Souza [1] are compared with those predicted according to Eq. (6.28). The data of Gunn and De Souza, shown in Figure 6.1, are those for the glass–air system with $k_S = 0.79$ W m^{-1} K^{-1} and $k_F = 0.026$ W m^{-1} K^{-1} (at an assumed temperature of 25°C). At $k_S/k_F = 30$, the effective thermal conductivity of the quiescent bed is found, in terms of k_e^o/k_F, from Figure 5.9, to be 6.1. With this value of k_e^o/k_F and $Pr = 0.7$, the $D_p U/\alpha_{ax}$-Re relationship predicted from Eq. (6.28) is shown as the lower solid line in Figure 6.1. As depicted, the experimental data match the theoretical values well, but differ considerably from those predicted according to Eq. (6.3), based on the mass dispersion analogy, over a range of Reynolds number from about 0.1 to 100. Incorporated in Figure 6.1 are some α_{ax} data determined from one-shot thermal response measurements by Gunn et al. [8]. These data, which were obtained from an analysis in the Laplace domain, also fit the theoretical line well. Wakao et al. [7, 9] also determined, from a time domain analysis of one-shot thermal response measurements, values of α_{ax} for air flow in beds of polystyrene foam particles, glass beads and lead shot. They reported that their results were correlated well by Eq. (6.28).

Corresponding to Eq. (6.27), the radial fluid thermal dispersion coefficient is given as:

$$\frac{\alpha_r}{\alpha_F} = \frac{k_{er}}{\epsilon_b k_F} \tag{6.29}$$

where k_{er} is the effective radial thermal conductivity.

6.4   Transient Effective Thermal Conductivities of Quiescent Beds

The axial thermal dispersion coefficient, α_{ax}, obtained from thermal response measurements is related, through Eq. (6.26), to the effective axial thermal conductivity, k_{eax}, which is usually determined under steady-state conditions. The purpose of this section is to discuss the effective thermal conductivity under unsteady-state conditions; a subject which has been examined by Kaguei et al. [10].

The conduction of heat in a cylindrical cell (radius R', length $2R$) shown in Figure 6.5 is considered. The system consists of a solid sphere (radius R, temperature T_S^*) and a stagnant fluid envelope (temperature T_F^*). For convenience, either end of the cylindrical cell is assumed to be at constant temperature. The temperatures, T_S^* and T_F^*, are then symmetric around the central axis of the cylinder.

The system is described by the following equations (in both cylindrical (r', x) and spherical (r, θ) coordinates):

$$\frac{\partial T^*}{\partial t} = \alpha \nabla^2 T^* \qquad \text{for } 0 < r' < R' \text{ and } -R < x < R \quad (6.30)$$

where

$$\nabla^2 = \frac{1}{r^2} \frac{\partial}{\partial r}\left(r^2 \frac{\partial}{\partial r}\right) + \frac{1}{r^2 \sin\theta} \frac{\partial}{\partial \theta}\left(\sin\theta \frac{\partial}{\partial \theta}\right)$$

$$T^* = T_S^* \text{ and } \alpha = \alpha_S \qquad \text{for } 0 < r < R$$

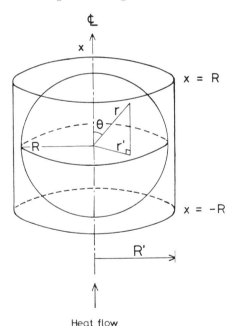

Heat flow

FIGURE 6.5   Heat transfer cell and coordinates.

and

$$T^* = T_F^* \text{ and } \alpha = \alpha_F \qquad \text{for } r > R$$

with

$$T_F^* = T_1 \qquad \text{at } x = -R$$

$$T_F^* = T_2 \qquad \text{at } x = R$$

and

$$\left.\begin{array}{l} T_F^* = T_S^* \\[2mm] k_F \dfrac{\partial T_F^*}{\partial r} = k_S \dfrac{\partial T_S^*}{\partial r} \end{array}\right\} \quad \text{at } r = R.$$

The rate of axial heat conduction through a cross-section (at x) of the cell is

$$q_x = -2\pi \int_0^{R'} k \frac{\partial T^*}{\partial x} r' \, dr' \tag{6.31}$$

where

$$k = k_S \text{ and } T^* = T_S^* \qquad \text{for } 0 < r < R$$

and

$$k = k_F \text{ and } T^* = T_F^* \qquad \text{for } r > R.$$

Therefore, the average axial heat conduction rate in the cell is

$$\bar{q} = \frac{1}{2R} \int_{-R}^{R} q_x \, dx. \tag{6.32}$$

The effective thermal conductivity, $K^o(t)$, may be defined as:

$$\bar{q} = K^o(t)\, \pi R'^2 \left(\frac{T_1 - T_2}{2R} \right) \tag{6.33}$$

or

$$K^o(t) = \frac{1}{\pi R'^2 (T_1 - T_2)} \int_{-R}^{R} q_x \, dx. \tag{6.34}$$

Suppose the cylinder ($R' = 1.05R$ and $\epsilon_b = 0.4$) shown in Figure 6.5 is heated under the conditions: $T^* = 0$ at $t = 0$; $T^* = 1$ at $x = -R\,(T_1 = 1)$ and $T^* = -1$ at $x = R$ (or $T_2 = -1$); $\partial T^*/\partial r' = 0$ at $r' = R'$. The temperature profiles in the lower half $(-R < x < 0)$ of the cylinder are computed with a grid network: the temperatures at the nodal points are calculated at sequences of time.

Figures 6.6(a) and (b) illustrate the increase in temperature in the cylinder, with a 1 mm glass sphere in it, at 0.001 and 0.09 s, respectively. The transient effective thermal conductivities are evaluated by the following rewritten form of Eq. (6.34):

$$K^o(t) = \frac{1}{\pi R'^2} \sum_{i,j} K_{i,j} [T^*_{i,j}(t) - T^*_{i+1,j}(t)] \, \Delta A_j \tag{6.35}$$

where, referring to Figure 6.6(a), $K_{i,j}$ is the rod conductance between nodes (i, j) and $(i + 1, j)$, and ΔA_j is the area represented by the rod at j. The transient effective thermal conductivity, $K^o(t)$, of the cylinder containing the 1 mm glass sphere is $0.2k_e^o$, at $t = 0.001$ s (Figure 6.6a); and $0.78k_e^o$, at $t = 0.09$ s (Figure 6.6b), where k_e^o is the steady-state effective thermal conductivity of a quiescent bed.

Figure 6.7 shows the conductivity–time curves for the cylinder containing a 1 mm glass/lead sphere and stagnant air. It is shown that the conductivities increase rapidly up to the steady-state values. The rise in conductivity for the lead–air system, is almost instantaneous, reaching its steady-state value in about 0.01 s. For the glass–air system, the increase is more gradual; but the transient time is still very short.

Since a packed bed may be visualized as composed of many unit cells connected in series, as discussed in Section 2.3, the conductivity of a

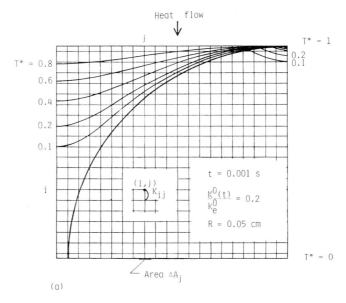

(a)

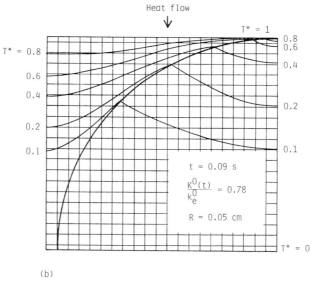

(b)

FIGURE 6.6   Grid network and computed temperature profiles, for glass (1 mm diameter)–air system: (a) $t = 0.001$ s; (b) $t = 0.09$ s.

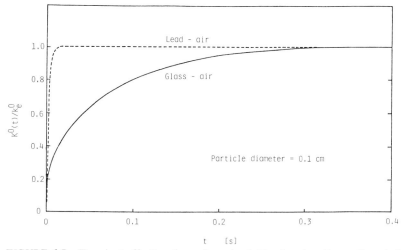

FIGURE 6.7   Transient effective thermal conductivities for glass (1 mm diameter)–
air and lead (1 mm diameter)–air systems.

quiescent bed is, thus, considered to be the same as that of the single cell
illustrated. When fluid is flowing in the bed, the conductivity is considered to attain its steady-state value in a much shorter time. In fact,
the transient time is generally very small compared to either the period of
frequency response (see Littman et al. [11] and Gunn and De Souza [1]),
or the residence time in an input–response measurement (Wakao [7]).
Hence, the effective thermal conductivity, defined by Eq. (6.33), may be
satisfactorily assumed to be constant and equal to its steady-state value, in
an overall unsteady-state heat transfer process.

6.5   Assumption of an Infinite Bed

The heat transfer parameters involved in Eqs. (6.1), (6.2) and (6.2a) based
on the D–C model have been determined by Gunn and De Souza [1] from
frequency response measurements, and by Gunn et al. [8] and Shen et al.
[4, 5] from one-shot response measurements. In the analysis, the packed
beds were assumed to be of infinite length. This assumption simplifies the
solution to Eqs. (6.1)–(6.2).

As in Section 1.4, it is necessary to examine where the response signal should be measured in a bed of finite length in order to satisfy the infinite bed assumption.

Let us assume that a finite packed bed is connected to an infinitely long empty column, as shown in Figure 1.26. Consider that a temperature change is imposed on a fluid flowing in the bed at $x < 0$, and the fluid temperature-time curves are monitored at $x = 0$ and $x = L$, which is at a distance, l, away from the bed exit.

According to the D–C model, the unsteady-state heat balance equations for the packed bed $(x < L + l)$ are given by Eqs. (6.1) and (6.2), and for the empty column $(x > L + l$; fluid temperature T'_F; fluid velocity u) by

$$\frac{\partial T'_F}{\partial t} = \frac{k'_F}{C_F \rho_F} \frac{\partial^2 T'_F}{\partial x^2} - u \frac{\partial T'_F}{\partial x}. \qquad (6.36)$$

The initial and boundary conditions are

$$T_F = T_S = T'_F = 0 \qquad\qquad \text{at } t = 0$$

$$T_F = T'_F \text{ and } k_{eax} \frac{\partial T_F}{\partial x} = k'_F \frac{\partial T'_F}{\partial x} \qquad \text{at } x = L + l$$

$$T'_F = 0 \qquad\qquad \text{at } x = \infty$$

where k'_F is the axial fluid thermal conductivity in flowing system, and k_{eax} is the effective axial thermal conductivity of the packed bed under unsteady-state conditions. As discussed in the preceding section, the unsteady-state conductivity is the same as that under steady-state conditions.

The transfer function of the system between the gas temperatures at $x = 0$ and $x = L$ is then

$$F(s) = \frac{1 - A_H \exp\left[-\sigma_H(l/L)\right]}{1 - A_H \exp\left\{-\sigma_H[1 + (l/L)]\right\}} \exp\left(\lambda_H\right) \qquad (6.37)$$

where

$$A_H = \frac{\lambda_H - \lambda'_H}{\lambda_H + \sigma_H - \lambda'_H} \qquad (6.37a)$$

$$\lambda_H = \frac{1}{2}\left(\frac{LU}{\alpha_{ax}} - \sigma_H\right) \tag{6.37b}$$

$$\sigma_H = \frac{LU}{\alpha_{ax}}(1 + B)^{1/2} \tag{6.37c}$$

$$\lambda'_H = \frac{LU}{2\alpha_{ax}}\left[1 - \left(1 + \frac{4k'_F s}{u^2 C_F \rho_F}\right)^{1/2}\right] \tag{6.37d}$$

and B is defined by Eq. (6.10a).

The first moment, M_1^{II}, of the impulse response is

$$M_1^{II} = -F'(0)$$

$$= \frac{L}{U}(1 + \beta_H)(1 - \Lambda_H) \tag{6.38}$$

where

$$\beta_H = \frac{aR}{3\epsilon_b}\frac{C_S \rho_S}{C_F \rho_F} \tag{6.38a}$$

$$\Lambda_H = N_H \Gamma_H \exp\left[-(l/L)/N_H\right][1 - \exp(-1/N_H)] \tag{6.38b}$$

$$N_H = \frac{\alpha_{ax}}{LU} \tag{6.38c}$$

and

$$\Gamma_H = 1 - \frac{k'_F}{\epsilon_b k_{eax}(1 + \beta_H)}. \tag{6.38d}$$

For an infinite bed ($l/L = \infty$), Eq. (6.38b) shows that $\Lambda_H = 0$; therefore, Λ_H is a measure of the deviation of the first moment from that of an infinite bed. In gas–solid systems, $\beta_H \gg 1$ and $k_{eax}/k'_F > 1$ and hence, Eq. (6.38d) shows that $\Gamma_H \simeq 1$ and Eq. (6.38b) reduces to

$$\Lambda_H \simeq N_H \exp\left[-(l/L)/N_H\right][1 - \exp(-1/N_H)]. \tag{6.39}$$

Based on Eq. (6.39), the relationships between Λ_H and N_H at values of l/L between 0 and 10 are presented in Figure 6.8. The criterion for an infinite bed may be given by $\Lambda_H = 0.01$. Thus, a packed bed with $N_H < 0.3$ may be assumed to be infinite if the response signal is measured at $l/L > 1$. If $N_H = 1$, however, the response signal should be measured at $l/L > 4$.

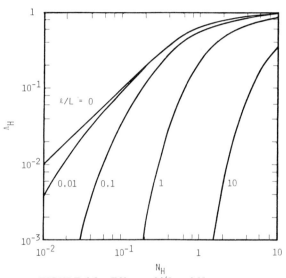

FIGURE 6.8   Effects of l/L and N_H on Λ_H.

With Eq. (6.28), the thermal dispersion number given by Eq. (6.38c) is rewritten as

$$N_H = \left[\frac{k_e^o}{k_F} \frac{1}{(Pr)(Re)} + 0.5 \right] \frac{D_p}{L}. \tag{6.40}$$

For most solid–gas packed bed systems, the ratio, k_e^o/k_F, is in the range 5 to 15, and $Pr \sim 1$. The condition $N_H < 0.3$ at $l/L > 1$ is met, therefore, at $Re > 2$ to 6 (corresponding to $k_e^o/k_F = 5$ to 15) for beds with $L/D_p = 10$. If $L/D_p = 20$, the condition is satisfied with $Re > 1$ to 3.

Shen et al. [4] applied the technique of a one-shot thermal input to the determination of the α_{ax}-Nu relation for the flow of air through a finite packed bed of glass beads (Example 6.1). Their measured input and

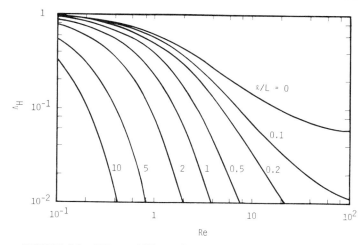

FIGURE 6.9   Effects of l/L and Re on Λ_H, for Run 2 in Example 6.1.

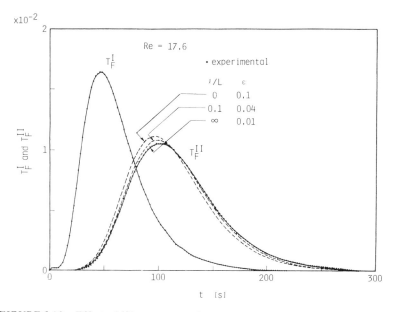

FIGURE 6.10   Effect of l/L on computed response signal, for Run 2 in Example 6.1.

response signals at $Re = 17.6$ are illustrated in Figure 6.2. Using their data ($L/D_p = 10$; $Pr = 0.7$; $k_e^0/k_F = 6.4$), the Λ_H-Re relationships at various l/L values are computed from Eqs. (6.39) and (6.40), and presented in Figure 6.9. The graph shows that at $Re = 17.6$, the condition, $\Lambda_H < 0.01$, is met, when l/L is greater than 0.3. In the experiments of Shen et al., the response signals were measured at $l/L \simeq 3$. The measured response signals are, therefore, free from the bed-end effect, and the assumption of an infinite bed is valid.

The response signals, $T_F^{II}(t)$, at $l/L = 0$, 0.1 and ∞ are computed using the heat transfer parameters ($\alpha_{ax} = 7 \times 10^{-4}$ m^2 s^{-1} and $Nu = 10$, which are the same as those employed in Figure 6.2), and compared, in Figure 6.10, with the response signal measured at $l/L \simeq 3$. As depicted, the deviation of the computed response curves from the measured signal increases with decreasing ratio of l/L (in the graph the deviation is indicated in terms of the root-mean-square-error, ϵ, defined by Eq. 6.18). Noticeable deviation is seen, when l/L is 0.1 or less.

Therefore, it may be concluded that the assumption of infinite bed is valid in beds of finite length, if response signals are measured at proper locations in the bed. Figure 6.8 demonstrates the locations where response signals should be measured in order to satisfy the criterion expressed in terms of Λ_H.

REFERENCES

[1]   D. J. Gunn and J. F. C. De Souza, *Chem. Eng. Sci.* **29**, 1363 (1974).
[2]   M. F. Edwards and J. F. Richardson, *Chem. Eng. Sci.* **23**, 109 (1968).
[3]   D. J. Gunn and C. Pryce, *Trans. Inst. Chem. Eng.* **47**, T341 (1969).
[4]   J. Shen, S. Kaguei and N. Wakao, *Chem. Eng. Sci.* **36**, 1283 (1981).
[5]   J. Shen, S. Kaguei and N. Wakao, *J. Chem. Eng. Japan* **14**, 413 (1981).
[6]   D. Vortmeyer, *Chem. Eng. Sci.* **30**, 999 (1975).
[7]   N. Wakao, *Chem. Eng. Sci.* **31**, 1115 (1976).
[8]   D. J. Gunn, P. V. Narayanan and A. P. Wardle, *Sixth Int. Heat Transfer Conf.* Toronto, Vol. 4, p. 19 (1978).
[9]   N. Wakao, S. Tanisho and B. Shiozawa, *Kagaku Kogaku Ronbunshu* **2**, 422 (1976).
[10] S. Kaguei, B. Shiozawa and N. Wakao, *Chem. Eng. Sci.* **32**, 507 (1977).
[11] H. Littman, R. G. Barile and A. H. Pulsifer, *Ind. Eng. Chem. Fund.* **7**, 554 (1968).

7 Unsteady-State Heat Transfer Models

THE THREE models proposed to describe the phenomena of heat transfer in packed beds under unsteady-state conditions are the Schumann model, the Continuous-Solid Phase model (C-S model) and the Dispersion-Concentric model (D-C model).

The Schumann model, the least complex of the three, is simply based on the assumption of ideal plug flow of the fluid and considers no heat conduction resistance in the solid particle.

The C-S model takes the solid phase heat conduction effect into consideration by assuming that the solid is in a continuous phase. Moreover, fluid thermal dispersion is also considered in the C-S model.

The D-C model is, by far, the most widely practised model in the solution of problems of unsteady-state heat transfer in packed beds. The model, which has already been discussed in Chapter 6, is based on the fluid having dispersed plug flow and the intraparticle temperature having radial symmetry. However, the assumption that the intraparticle temperature profile is radially symmetric does not portray the real temperature profile in the solid particle. Similar to the discussion in Section 1.2.4, if the temperature profile in particle was concentric, no heat conduction would take place across the particle. To compensate for this shortcoming, solid phase heat conduction has to be superficially included in the fluid thermal dispersion term. Solid phase conduction is important in unsteady-state heat transfer, particularly at low flow rates.

In this chapter, the unsteady-state heat transfer models are discussed and compared with respect to their response signals. The Nusselt numbers, Nu, predicted from the corresponding response signals, are assessed to determine the validity of the models. The underlying assumption of an intraparticle concentric temperature profile in the D-C model is also verified.

243

7.1   Step and Frequency Responses for the Schumann, C–S and D–C Models

Equations (7.1)–(7.7) listed in Table 7.1 are the fundamental equations of the Schumann, C–S and D–C models for unsteady-state heat transfer in inert packed beds. Besides the notation employed in Chapter 6, the following symbols are used (in the C–S model):

$$k_{eF} = \text{effective fluid phase thermal conductivity}$$

$$k_{eS} = \text{effective solid phase thermal conductivity}$$

7.1.1.   Transfer Functions for Response Signals at Bed Exit

Transfer functions governing the thermal input signal, $T_F^I(t)$, imposed on a fluid entering the bed, and the response signal, $T_F^{II}(t)$, of a fluid leaving a bed of length, L, are shown below:

7.1.1.1   *The Schumann model*

The boundary condition for the Schumann model [3] is

$$T_F = T_F^I(t) \qquad \text{at } x = 0 \text{ (inlet).} \tag{7.8}$$

The transfer function, relating the input and response signals, is

$$[F(s)]_{\text{Schumann}} = \exp\left[-\frac{sL}{U}\left(1 + \frac{k_1}{s + k_2}\right)\right] \tag{7.9}$$

where

$$k_1 = \frac{h_p a}{\epsilon_b C_F \rho_F} \tag{7.9a}$$

and

$$k_2 = \frac{h_p a}{(1 - \epsilon_b) C_S \rho_S}. \tag{7.9b}$$

7.1.1.2   *The C–S model*

The boundary conditions employed by Littman *et al.* [1, 2] are

TABLE 7.1

Heat transfer models and the fundamental equations.

Model	Assumptions	Fundamental equations	
Schumann model	Fluid in plug flow	$\dfrac{\partial T_F}{\partial t} = -U\dfrac{\partial T_F}{\partial x} - \dfrac{h_p a}{\epsilon_b C_F \rho_F}(T_F - T_S)$	(7.1)
	No temperature gradient within the particle	$(1-\epsilon_b)\dfrac{\partial T_S}{\partial t} = \dfrac{h_p a}{C_S \rho_S}(T_F - T_S)$	(7.2)
C–S model	Fluid in dispersed plug flow	$\dfrac{\partial T_F}{\partial t} = \dfrac{k_{eF}}{\epsilon_b C_F \rho_F}\dfrac{\partial^2 T_F}{\partial x^2} - U\dfrac{\partial T_F}{\partial x} - \dfrac{h_p a}{\epsilon_b C_F \rho_F}(T_F - T_S)$	(7.3)
	Axial heat conduction in the solid phase	$(1-\epsilon_b)\dfrac{\partial T_S}{\partial t} = \dfrac{k_{eS}}{C_S \rho_S}\dfrac{\partial^2 T_S}{\partial x^2} + \dfrac{h_p a}{C_S \rho_S}(T_F - T_S)$	(7.4)
D–C model	Fluid in dispersed plug flow Dispersion coefficient of Eq. (6.3) for the original D–C model; that of Eq. (6.28) for the modified D–C model	$\dfrac{\partial T_F}{\partial t} = \alpha_{ax}\dfrac{\partial^2 T_F}{\partial x^2} - U\dfrac{\partial T_F}{\partial x} - \dfrac{h_p a}{\epsilon_b C_F \rho_F}|T_F - (T_S)_R|$	(7.5)
	Particle temperature with radial symmetry	$\dfrac{\partial T_S}{\partial t} = \alpha_S \dfrac{1}{r^2}\dfrac{\partial}{\partial r}\left(r^2 \dfrac{\partial T_S}{\partial r}\right)$	(7.6)
		$k_S\left(\dfrac{\partial T_S}{\partial r}\right) = h_p(T_F - T_S) \quad \text{at } r = R$	(7.7)

original

modified

In the C–S model, k_{eF} = effective fluid phase thermal conductivity and k_{eS} = effective solid phase thermal conductivity, both based on the bed cross-section. The same symbols are used by Littman et al. [1, 2], but $k_{eF} = \epsilon_b (k_{eF})$Littman and $k_{eS} = (1-\epsilon_b)(k_{eS})$Littman.

$$\left.\begin{array}{l} \epsilon_b C_F \rho_F U [T_F - T_F^I(t)] = k_{eF} \dfrac{\partial T_F}{\partial x} \\[2mm] -k_{eS} \dfrac{\partial T_S}{\partial x} = (1 - \epsilon_b) h_p (T_F - T_S) \end{array}\right\} \text{at } x = 0 \text{ (inlet)}$$

$$\left.\begin{array}{l} \dfrac{\partial T_F}{\partial x} = 0 \\[2mm] k_{eS} \dfrac{\partial T_S}{\partial x} = (1 - \epsilon_b) h_p (T_F - T_S) \end{array}\right\} \text{at } x = L \text{ (exit).}$$

$$(7.10)$$

The transfer function is

$$[F(s)]_{C-S} = \sum_{i=1}^{4} P_i \exp(m_i) \qquad (7.11)$$

where m_i is an i-th root of the following equation:

$$m_i^4 - \frac{\epsilon_b C_F \rho_F UL}{k_{eF}} m_i^3 - \left[\frac{\epsilon_b C_F \rho_F s + h_p a}{k_{eF}} + \frac{(1 - \epsilon_b) C_S \rho_S s + h_p a}{k_{eS}} \right] L^2 m_i^2$$

$$+ \frac{\epsilon_b C_F \rho_F UL^3}{k_{eF} k_{eS}} [(1 - \epsilon_b) C_S \rho_S s + h_p a] m_i + \left[\frac{\epsilon_b (1 - \epsilon_b) C_F \rho_F C_S \rho_S}{k_{eF} k_{eS}} s^2 \right.$$

$$\left. + \frac{\epsilon_b C_F \rho_F + (1 - \epsilon_b) C_S \rho_S}{k_{eF} k_{eS}} h_p a s \right] L^4 = 0 \qquad (7.12)$$

and P_i is an i-th root of

$$\left.\begin{array}{l} \displaystyle\sum_{i=1}^{4} \left(1 - \dfrac{k_{eF}}{\epsilon_b C_F \rho_F UL} m_i \right) P_i = 1 \\[4mm] \displaystyle\sum_{i=1}^{4} \left(v_i + \dfrac{h_p a L^2}{1 - w_i} \right) P_i = 0 \end{array}\right\}$$

$$\sum_{i=1}^{4} m_i \exp{(m_i)} P_i = 0$$

$$\sum_{i=1}^{4} \left(v_i + \frac{h_p a L^2}{1+w_i}\right) \exp{(m_i)} P_i = 0 \qquad (7.13)$$

in which

$$v_i = k_{eF} m_i^2 - \epsilon_b C_F \rho_F UL \left[m_i + \left(s + \frac{h_p a}{\epsilon_b C_F \rho_F}\right)\frac{L}{U}\right]$$

$$w_i = \frac{k_{eS} m_i}{(1-\epsilon_b) h_p L} \; .$$

7.1.1.3   The D-C model

Assuming that the Danckwerts boundary conditions [4] may be applied to the fundamental heat balance equations. Then,

$$U[T_F - T_F^I(t)] = \alpha_{ax} \frac{\partial T_F}{\partial x} \qquad \text{at } x = 0 \text{ (inlet)}$$

$$\frac{\partial T_F}{\partial x} = 0 \qquad \text{at } x = L \text{ (exit).} \qquad (7.14)$$

The transfer function is given as:

$$[F(s)]_{\text{D-C}} = \frac{4A \exp\left(\dfrac{LU}{2\alpha_{ax}}\right)}{(1+A)^2 \exp\left(A \dfrac{LU}{2\alpha_{ax}}\right) - (1-A)^2 \exp\left(-A \dfrac{LU}{2\alpha_{ax}}\right)} \qquad (7.15)$$

where

$$A = (1+B)^{1/2} \qquad (7.15a)$$

and B is defined in Eq. (6.10a). Note that $\alpha_S = k_S/(C_S \rho_S)$.

7.1.2   Prediction of Step and Frequency Responses

7.1.2.1   *Step response*

With the following conditions

$$T_F = T_S = 0 \qquad \text{at } t = 0$$
$$T_F^I(t) = T_{in} \qquad \text{at } t > 0 \tag{7.16}$$

the response, $T_F^{II}(t)$, is obtained as:

$$\frac{T_F^{II}(t)}{T_{in}} = \frac{1}{2} + \frac{2}{\pi} \sum_{n=1}^{\infty} \frac{1}{2n-1} \, \text{Imag}[F(s) \exp(st)]_{s=i(2n-1)\pi/\tau^*} \tag{7.17}$$

where τ^* is a time sufficiently long enough to allow the step response to attain a steady-state value.

7.1.2.2   *Frequency response*

The response, $T_F^{II}(t)$, under stationary conditions, is obtained using

$$T_F^I(t) = A_\omega^I \cos \omega t \tag{7.18}$$

as

$$\frac{T_F^{II}(t)}{A_\omega^I} = \text{Real}[F(s) \exp(st)]_{s=i\omega} \tag{7.19}$$

where A_ω^I is the amplitude of the harmonic component with frequency ω of the input signal.

When the responses are measured at the bed exit, the transfer functions, $F(s)$ in Eqs. (7.17) and (7.19) are substituted by Eqs. (7.9), (7.11) and (7.15) for the Schumann, C–S and D–C models, accordingly.

7.2   Assumption of a Concentric Temperature Profile in a Solid Sphere in the D–C Model

The fundamental equations, Eqs. (7.5)–(7.7), based on the assumptions of dispersed plug flow and concentric intraparticle temperature profiles, are rather simple and can easily be solved. The question is whether the assumption of intraparticle temperature being radially symmetric can be verified in the calculation of the heat transfer rate. The purpose of this section is to examine the validity of this assumption.

For a particle in a packed bed, the unsteady-state heat balance equation is given in terms of temperature T_S^* as:

$$\frac{\partial T_S^*}{\partial t} = \alpha_S \nabla^2 T_S^* \tag{7.20}$$

where

$$\nabla^2 = \frac{1}{r^2} \frac{\partial}{\partial r}\left(r^2 \frac{\partial}{\partial r}\right) + \frac{1}{r^2}\left[\frac{1}{\sin\theta}\frac{\partial}{\partial\theta}\left(\sin\theta\frac{\partial}{\partial\theta}\right) + \frac{1}{\sin^2\theta}\frac{\partial^2}{\partial\Phi^2}\right] \cdot \tag{7.20a}$$

Following the same procedure given in Section 1.2.4, Eq. (7.20) is replaced by

$$\frac{1}{4\pi r^2} \int_0^{2\pi} d\Phi \int_{-1}^{1} \frac{\partial T_S^*}{\partial t} r^2 \, d\cos\theta = \frac{\alpha_S}{4\pi r^2} \int_0^{2\pi} d\Phi \int_{-1}^{1} \nabla^2 T_S^* r^2 \, d\cos\theta . \tag{7.21}$$

Changing the order of differentiation and integration and then taking $(\partial T_S^*/\partial\Phi)_{\Phi=0} = (\partial T_S^*/\partial\Phi)_{\Phi=2\pi}$ into consideration, it results in the following expression:

$$\frac{\partial X}{\partial t} = \alpha_S \frac{1}{r^2} \frac{\partial}{\partial r}\left(r^2 \frac{\partial X}{\partial r}\right) \tag{7.22}$$

where

$$X = \frac{1}{4\pi} \int_0^{2\pi} d\Phi \int_{-1}^{1} T_S^* \, d\cos\theta . \tag{7.22a}$$

This X is nothing but the average value of T_S^* on a spherical surface with radius r. The actual temperatures, T_S^*, in a particle are not radially

symmetric, but the intraparticle temperature profile, similar to the intraparticle concentration profile argument, can be expressed in terms of
average temperatures which are radially symmetric.

The rate of heat transfer from a fluid to a particle is then

$$Q_p(t) = k_S \left(\int_0^{2\pi} d\Phi \int_{-1}^{1} \frac{\partial T_S^*}{\partial r} R^2 \, d\cos\theta \right)_R$$

$$= k_S R^2 \left(\frac{\partial}{\partial r} \int_0^{2\pi} d\Phi \int_{-1}^{1} T_S^* \, d\cos\theta \right)_R. \qquad (7.23)$$

Substitution of Eq. (7.22a) into Eq. (7.23) gives

$$Q_p(t) = 4\pi R^2 k_S \left(\frac{\partial X}{\partial r} \right)_{r=R}. \qquad (7.24)$$

Therefore, we may conclude that the heat transfer rate, $Q_p(t)$, may be
evaluated in terms of the intraparticle surface mean temperature.

7.3   Effect of Fluid Thermal Dispersion Coefficients on the Nusselt Numbers of the D–C model

The axial solid phase heat conduction contribution is not taken into consideration in the D–C model as originally proposed. To overcome this
problem, Wakao [5] proposed including it in the fluid dispersion term. He
has shown that the axial fluid thermal dispersion coefficient, α_{ax}, is related
to the effective axial thermal conductivity, k_{eax}, according to Eq. (6.26) or
Eq. (6.28). The model, which takes into account axial solid phase heat
conduction, is henceforth referred to as the modified D–C model in order
to distinguish it from the original D–C model in which thermal dispersion
is simply based on a mass dispersion analogy.

The effect of fluid thermal dispersion coefficients on the particle-to-
fluid heat transfer coefficients based on the two D–C models will be
examined using the following example:

Example 7.1

Let us consider that the thermal frequency response signals are measured over a distance, $L = 1.5$ cm, in an infinitely long column packed with glass beads (1 mm) and bed void fraction, $\epsilon_b = 0.4$. Air is flowing in the bed at $Re < 2$. Additional data are: thermal conductivities, $k_F = 0.027$ and $k_S = 0.88$ W m^{-1} K^{-1}; specific heats, $C_F = 1000$ and $C_S = 670$ J kg^{-1} K^{-1}; densities, $\rho_F = 1.2$ and $\rho_S = 2500$ kg m^{-3}. Find Nusselt numbers based on the original D–C model.

SOLUTION

The frequency response signals, based on the D–C models, should be computed using Eqs. (6.10) and (7.19) from the above information together with axial thermal dispersion coefficients and particle-to-fluid heat transfer coefficients. The frequency response signal based on the modified D–C model is denoted by $(T_F^{II})_{D-C}$. This is computed using α_{ax} from Eq. (6.28) and values of Nusselt number, Nu, ranging from 1 to ∞. (Note that $(T_F^{II})_{D-C}$ is not influenced by Nu in the range 1 to ∞.) Similarly, the frequency response signal, based on the original D–C model, is calculated using α'_{ax} from Eq. (6.3) and various assumed values of Nu'. This signal is denoted by $(T_F^{II})'_{D-C}$.

The correct values of Nu' are then determined by minimizing the difference between the two calculated response signals using the root-mean-square-error, ϵ_f, defined as:

$$\epsilon_f = \left\{ \frac{\displaystyle\int_0^{2\pi/\omega} [(T_F^{II})_{D-C} - (T_F^{II})'_{D-C}]^2 \, dt}{\displaystyle\int_0^{2\pi/\omega} [(T_F^{II})_{D-C}]^2 \, dt} \right\}^{1/2}. \tag{7.25}$$

Figure 7.1 compares the two frequency response signals, $(T_F^{II})_{D-C}$ and $(T_F^{II})'_{D-C}$, computed at $\omega = 0.0004\pi$ rad s^{-1} and $Re = 2$. We can say that the two signals are in:

a)   Very good agreement, if $\epsilon_f < 0.01$.

b)   Relatively good agreement, if $0.01 < \epsilon_f < 0.05$.

c)   Poor agreement, if $\epsilon_f \gg 0.05$.

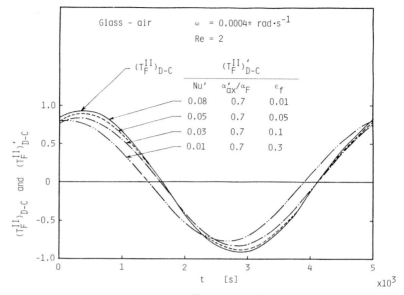

FIGURE 7.1   Comparison of $(T_F^{II})_{D-C}$ and $(T_F^{II})'_{D-C}$ for Example 7.1.

Figure 7.2 is the error map of Nu' versus Re at a frequency, $\omega = 0.0001\pi$ rad s^{-1}. Based on the criteria, (a)–(c) above, the following results may be drawn:

a') At $Re = 2$, any value of Nu' greater than about 0.03 gives very good agreement ($\epsilon_f < 0.01$), therefore, no definite Nu' value can be determined.

b') At $Re = 0.9$ with $\epsilon_f = 0.01$ (Point A), Nu' is found to be 0.015; similarly, at $Re = 0.5$ with $\epsilon_f = 0.05$ (Point B), Nu' is shown to be 0.008.

c') For $Re < 0.5$, the fit is always poor ($\epsilon_f \gg 0.05$) with any value of Nu'.

Therefore, it can be concluded that, within an acceptable confidence range, curve AB is the only Nu'-Re relationship to be found from thermal response at this frequency.

Similar Nu-Re relationships for a number of different systems with varied response frequencies are estimated and presented in Figure 7.3. As

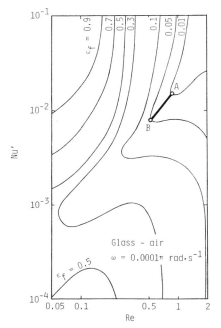

FIGURE 7.2   Error map in the plot of Nu' versus Re for Example 7.1.

shown, each response frequency will predict only a certain range of the Nu'-Re relationship. This frequency dependency had been experimentally observed by Turner and Otten [6] and Gunn and De Souza [7].

If, in Figure 7.3, the Nu'-Re relationships for the various systems are interpreted as Nu'-$Pr\,Re$, then Figure 7.4 shows that all the data fall almost within the same range bounded by the two curves. At high Reynolds number, the mixing terms in Eqs. (6.3) and (6.28) predominate, so that the difference between α'_{ax} and α_{ax} is small; hence, the two Nusselt values predicted from the original and modified D-C models should be close. At low Reynolds number, as shown in Figure 7.4 ($Re < 2$), the original D-C model which uses α'_{ax} from Eq. (6.3) is found to give entirely different and anomalously low Nusselt values.   *(End of Example)*

7.4   The C-S Model

Based on the assumption that the bed of solid particles is in a continuous phase, the C-S model, proposed by Littman and Barile [1], takes into

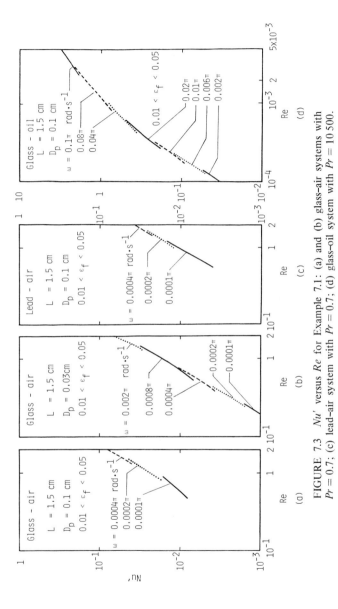

FIGURE 7.3 Nu' versus Re for Example 7.1: (a) and (b) glass–air systems with $Pr = 0.7$; (c) lead–air system with $Pr = 0.7$; (d) glass–oil system with $Pr = 10\,500$.

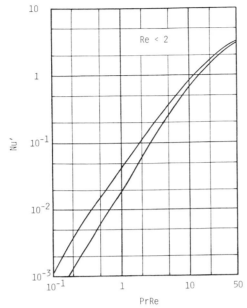

FIGURE 7.4 Nu' versus $(Pr)(Re)$ for Example 7.1.

consideration the solid phase heat conduction in addition to the fluid phase thermal dispersion. In this section, the C–S model will be examined by assessing the heat transfer parameters obtained from the model with respect to the equivalent parameters according to the modified D–C model.

The transfer functions and the frequency responses, given in Section 7.1, will be used in the analysis. The Nusselt numbers of these two models are denoted as Nu and Nu'' for the modified D–C model and the C–S model, respectively.

7.4.1   From the C–S Model to the Modified D–C Model

Based on the C–S model and thermal frequency response measurements, Littman *et al.* [2] obtained particle-to-fluid heat transfer coefficients in terms of Nusselt numbers, Nu'', and effective thermal conductivities of the solid phase, k_{eS}, for several solid–gas systems. Based on their results for

k_{eS} and Nu'', the heat transfer parameters, α_{ax} and Nu, can be predicted from the D-C model. The method of converting rate parameters of one model into the corresponding parameters of another model and vice versa is given by Kaguei *et al.* [8].

Using the data given by Littman *et al.* [2] and by assuming several frequency values, response signals, $(T_F^{II})_{C-S}$, based on the C-S model may be evaluated from Eqs. (7.11) and (7.19). As an illustration, let us consider their data for Run No. 28 (copper-air, $D_p = 0.72$ mm, $Re = 2.74$, $L = 1.32$ cm, $\epsilon_b = 0.46$), in which the Nusselt number, Nu'', and the effective solid phase thermal conductivity, k_{eS}, are reported to be 1.0 and 0.45 W m^{-1} K^{-1}, respectively. Using the given data, the amplitude ratio–frequency and the phase lag–frequency relationships at a Reynolds number of 2.74 are estimated as shown in Figure 7.5. The response signals, $(T_F^{II})_{C-S}$, are then calculated over the amplitude ratio 0.04 to 0.5 or the frequency range 0.0021π to 0.015π rad s^{-1}. On the other hand, thermal response signals based on the D-C model, $(T_F^{II})_{D-C}$, over the same frequency range and at the same Reynolds number can also be computed from Eqs. (7.15) and (7.19) with various assumed values of α_{ax} and Nu.

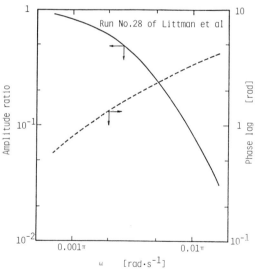

FIGURE 7.5   Amplitude ratio versus frequency, and phase lag versus frequency computed for Run No. 28 ($Re = 2.74$) of Littman *et al.* [2].

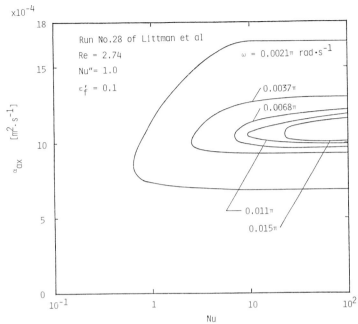

FIGURE 7.6 α_{ax}–Nu relationship for the D–C model evaluated using data of Run No. 28 ($Re = 2.74$) of Littman et al. [2].

From the two sets of response signals, $(T_F^{II})_{C-S}$ and $(T_F^{II})_{D-C}$, the α_{ax}–Nu relationship of the D–C model is determined by curve fitting. Figure 7.6 depicts such a relationship with frequency as a parameter at a Reynolds number of 2.74. The contour map is constructed with a root-mean-square-error, ϵ_f', defined by Eq. (7.26), of less than 0.1.

$$\epsilon_f' = \left\{ \frac{\displaystyle\int_0^{2\pi/\omega} [(T_F^{II})_{C-S} - (T_F^{II})_{D-C}]^2 \, dt}{\displaystyle\int_0^{2\pi/\omega} [(T_F^{II})_{C-S}]^2 \, dt} \right\}^{1/2}. \tag{7.26}$$

From the contour map with $\omega = 0.015\pi$ rad s^{-1}, α_{ax} is determined to be approximately 11×10^{-4} m^2 s^{-1}, which agrees well with the value, 11.2×10^{-4} m^2 s^{-1}, predicted from Eq. (6.28). However, the contour map indicates that the Nusselt value cannot possibly be determined according to the D–C model with α_{ax} from Eq. (6.28), i.e. the modified D–C model. At this low Reynolds number, the large and frequency-dependent confidence range of Nusselt number according to the modified D–C model indicates that the Nusselt number, Nu, may be regarded even as large as infinity, or there is little resistance to particle-to-fluid heat transfer. However, according to the C–S model, the particle-to-fluid heat transfer coefficient is given as $Nu'' = 1$.

7.4.2   From the Modified D–C Model to the C–S Model

The preceding section has demonstrated that the exact value of Nusselt number based on the modified D–C model cannot be ascertained from heat transfer parameters obtained according to the C–S model. The purpose of this section is to reverse the procedure and to see what the C–S model Nusselt numbers, Nu'', are generated from those based on the modified D–C model, Nu. The data for the glass–air system given in Example 7.1 are used in this illustration as follows:

The frequency response signals based on the models, $(T_F^{II})_{D-C}$ and $(T_F^{II})_{C-S}$ will be predicted at Reynolds numbers of 0.1 and 1. Using $Nu \geqslant 1$ and α_{ax} estimated from Eq. (6.28), $(T_F^{II})_{D-C}$ values can be evaluated from Eqs. (7.15) and (7.19) over the frequency range 0.0002π to 0.002π rad s^{-1} for $Re = 0.1$, and 0.002π to 0.01π rad s^{-1} for $Re = 1$. Note that any Nusselt number greater than unity will do; it makes no difference to the calculated response signals, $(T_F^{II})_{D-C}$. $(T_F^{II})_{C-S}$ values can be computed using Eqs. (7.11) and (7.19) with various assumed values of k_{eS} and Nu'' at the same Reynolds numbers. In the computation, the values of k_{eF} in Eqs. (7.12) and (7.13) are estimated using the following relationships:

$$k_{eF} = 0.7\epsilon_b k_F \qquad \text{at } Re \leqslant 0.8 \qquad (7.27)$$

$$= 0.5(Pr)(Re) k_F \qquad \text{at } Re > 0.8. \qquad (7.28)$$

A Reynolds number of 0.8 is tentatively chosen as the borderline in the application of Eqs. (7.27) and (7.28) with $\epsilon_b = 0.4$ and $Pr = 0.7$.

From the fitting of $(T_F^{II})_{C-S}$ and $(T_F^{II})_{D-C}$, the k_{eS}–Nu'' relationship according to the C–S model can be determined. Figures 7.7(a) and (b)

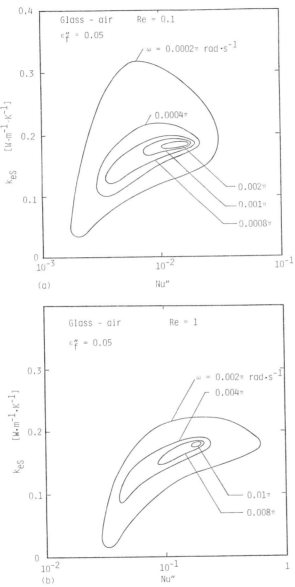

FIGURE 7.7 $k_{eS}\text{-}Nu''$ relationship for the C–S model, amplitude ratio 0.05 to 0.7:
(a) $Re = 0.1$; (b) $Re = 1$.

show the relationship with a root-mean-square-error, ϵ_f'', defined by Eq. (7.29), of 0.05.

$$\epsilon_f'' = \left\{ \frac{\displaystyle\int_0^{2\pi/\omega} [(T_F^{II})_{D\text{-}C} - (T_F^{II})_{C\text{-}S}]^2 \, dt}{\displaystyle\int_0^{2\pi/\omega} [(T_F^{II})_{D\text{-}C}]^2 \, dt} \right\}^{1/2} . \tag{7.29}$$

Littman *et al.* [2] observed experimentally that the k_{eS} values predicted according to the C–S model depended upon the frequency values applied. However, k_{eS} cannot be a function of frequency. As illustrated in Figures 7.7(a) and (b), it is only the confidence range of the k_{eS}–Nu'' relationship that depends on the frequency value.

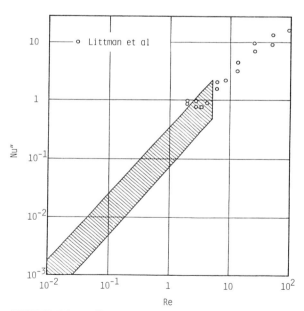

FIGURE 7.8 Nu'' versus Re relationship for the C–S model.

Figure 7.8 shows Nu'' predicted in the amplitude ratio range 0.04 to 0.5 for the glass–air and copper–air systems over a range of Reynolds numbers from 0.01 to 5. The numerous Nu'' data (within the shaded area) obtained over this range of Reynolds numbers are compared with the experimental data determined by Littman et al. As illustrated, the predicted values agree reasonably well with the experimental data in the upper Reynolds number range. At low flow rates, the estimated Nusselt numbers, Nu'', are significantly lower than the corresponding values determined according to Eq. (8.20); moreover, the continuous decrease in Nu'' with decreasing flow rate predicted by the C–S model is, as discussed in Chapter 8, illogical and contradictory. The anomaly appears to indicate that the C–S model does not depict satisfactorily the phenomena of heat transfer in packed beds.

7.5   The Schumann Model

The Schumann model [3] is based on the assumptions of fluid plug flow with no dispersion and no temperature gradient existing in the solid particle. In this section, the relationship between the Nusselt number, Nu, of the modified D–C model and Nu''' of the Schumann model is examined.

From step response measurements Handley and Heggs [9] obtained heat transfer coefficients based on the Schumann model. Using the information obtained from their work on the solid–gas system, the response curve, $(T_F^{II})_{Schumann}$, is predicted from Eqs. (7.9) and (7.17) based on the Schumann model. The $(T_F^{II})_{D-C}$ curves based on the modified D–C model are also computed from Eqs. (7.15) and (7.17) with various assumed values of the heat transfer coefficient. Figure 7.9 shows a comparison of the $(T_F^{II})_{Schumann}$ and $(T_F^{II})_{D-C}$ curves. The error, ϵ_s, is evaluated from

$$\epsilon_s = \left\{ \frac{1}{t_2 - t_1} \int_{t_1}^{t_2} \left[\frac{(T_F^{II})_{Schumann} - (T_F^{II})_{D-C}}{T_{in}} \right]^2 dt \right\}^{1/2} \tag{7.30}$$

where t_1 and t_2 are chosen for

$$[T_F^{II}(t_1)]_{Schumann} = 0.05 T_{in}$$

and

$$[T_F^{II}(t_2)]_{Schumann} = 0.95 T_{in}$$

respectively.

From Figure 7.9 it appears that the agreement between the two curves is good when $\epsilon_s < 0.02$. In Figure 7.10, the original Nusselt numbers, Nu''', of Handley and Heggs [9] are compared with those re-evaluated according to the modified D–C model. Some of the recalculated data are plotted with a confidence range indicating that $\epsilon_s = 0.02$. It is seen that the data re-evaluated according to the modified D–C model are generally higher than the corresponding values based on the Schumann model. The difference between the two Nusselt numbers, Nu and Nu''', appears to widen with a lowering of the Reynolds number.

From the analyses, it may be concluded that the Schumann model, which suffers from its over-simplified assumptions, is the least reliable for predicting heat transfer parameters. Both the C–S model and the original D–C model, although applicable at high Reynolds number, are found to give erroneously low and anomalous heat transfer coefficients at low Reynolds number. The modified D–C model, which also takes into consideration the effect of axial heat conduction in solid particle and uses Eq. (6.28) in the prediction of the axial fluid thermal dispersion coefficient, describes the phenomena of heat transfer in packed beds more closely.

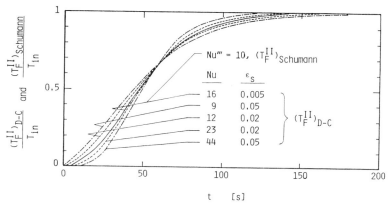

FIGURE 7.9   Comparison of step response curves, lead–air system; $Re \simeq 100$, $\epsilon_b = 0.36, D_p = 3$ mm, $L = 3.3$ cm (these data are from Handley and Heggs [9]).

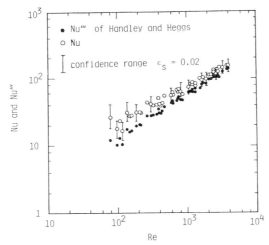

FIGURE 7.10 Nu''' data of Handley and Heggs [9] and Nu re-evaluated based on the modified D–C model.

REFERENCES

[1] H. Littman and R. G. Barile, *Chem. Eng. Prog. Symp. Ser.* **62** (No. 67), 10 (1966).
[2] H. Littman, R. G. Barile and A. H. Pulsifer, *Ind. Eng. Chem. Fund.* **7**, 554 (1968).
[3] T. E. W. Schumann, *J. Franklin Inst.* **208**, 405 (1929).
[4] P. V. Danckwerts, *Chem. Eng. Sci.* **2**, 1 (1953).
[5] N. Wakao, *Chem. Eng. Sci.* **31**, 1115 (1976).
[6] G. A. Turner and L. Otten, *Ind. Eng. Chem. Process Des. Dev.* **12**, 417 (1973).
[7] D. J. Gunn and J. F. C. De Souza, *Chem. Eng. Sci.* **29**, 1363 (1974).
[8] S. Kaguei, B. Shiozawa and N. Wakao, *Chem. Eng. Sci.* **32**, 507 (1977).
[9] D. Handley and P. J. Heggs, *Trans. Inst. Chem. Eng.* **46**, T251 (1968).

8 Particle-to-Fluid Heat Transfer Coefficients

FOR THE design and analysis of packed bed catalytic reactors, it is necessary to know the temperatures of the fluid and the catalyst particles in which the chemical reactions are taking place. In general, fluid temperature is measured with little difficulty, but the measurement of the solid surface temperature is not easy. This is particularly true of packed bed reactors. The particle temperature or temperature drop at the particle surface then has to be estimated in terms of the heat transfer coefficient between the particle and the fluid.

Because of the importance of the particle-to-fluid heat transfer coefficient in packed bed reactor, a considerable effort has been made to evaluate this parameter. Experimental determinations of heat transfer coefficients for a wide variety of systems have been made using various experimental techniques, under either steady-state or unsteady-state conditions. Table 8.1 summarizes, in chronological order, the heat transfer work reported in the literature together with their corresponding experimental methods and operating conditions.

Figure 8.1 compares the numerous experimental data on heat transfer coefficients published in the literature over a wide range of Reynolds numbers. The transfer coefficients are expressed in terms of Nusselt numbers, Nu, in the graph. As shown, the reported data at high Reynolds numbers are consistent; however, at low flow rates, the data are quite incompatible and indicate a continuous decrease in Nusselt number with decreasing Reynolds number. This anomalous decrease in Nusselt number at low Reynolds numbers has been the subject of long dispute. There are two opposing views: one supports the unlimited decline in Nusselt number based on experimental observations, whereas the other argues that a limiting Nusselt number should exist at zero flow rate. Gunn and De Souza [23], for example, have shown from frequency response measurements, that the limiting Nusselt number is about 10.

264

Some theoretical studies have also been carried out to explain the observed experimental results. There is, of course, no exact theory which describes satisfactorily the transport phenomena in packed beds. The proposed models which are based on different assumptions often predict different and contradicting Nusselt numbers. Two completely divergent conclusions have been drawn based on different assumptions.

From an analogy between electrostatics and heat, Cornish [31] explained that the heat transfer coefficient in a dense system of particles was considerably less than that for a single sphere in an infinite medium. Kunii and Suzuki [32] pointed out that fluid flowing in channels in a bed was the reason for this anomaly. Nelson and Galloway [33] claimed that the anomaly could be explained by a renewal of the fluid element surrounding each particle. Schlünder [34] showed that if a packed bed was a bundle of parallel capillaries the transfer coefficient should continuously decrease with a decrease in the flow rate at lower Reynolds numbers. Martin [35] pointed out that non-uniform packing of the particles in a bed caused a decrease in Nusselt number at low flow rates.

On the other hand, by applying a free surface model, Pfeffer and Happel [36] obtained a Nu–Re relationship with a limiting Nusselt number of about 13, as the Reynolds number dropped to zero in a bed with a void fraction of 0.4. From an analysis of steady-state heat transfer in a concentric hollow sphere with a stagnant fluid, Miyauchi [37] showed a limiting Nusselt number of about 18, at a bed void fraction of 0.4. Sørensen and Stewart [38] studied creep flow through a cubic array of spheres and found a limiting Nusselt number of about 3.9. Schlünder [39] proposed that the Nusselt numbers for particles in packed beds are greater, by a factor of $1 + 1.5(1 - \epsilon_b)$, than those for flow over single spheres; assuming this, he obtained a limiting Nusselt number of 3.8 for a packed bed with $\epsilon_b = 0.4$. Based on a stochastic model of extraparticle void space, Gunn [40] derived a theoretical equation for the particle-to-fluid transfer coefficient, which yields a limiting Nusselt number of 4 at zero flow rate.

Wakao *et al.* [29, 30, 41, 42], however, have shown that it is a defect in the fundamental equations which is mainly responsible for the anomalous decrease in Nusselt number at lower Reynolds numbers.

In the following sections a critical review is made of the published particle-to-fluid heat transfer coefficient data. Selected reliable data will be revised and correlated to give an empirical formula which may be used for the accurate prediction of particle-to-fluid heat transfer coefficients.

TABLE 8.1
Heat transfer experimental data[a].

Year	Investigator	Experimental method	Steady or unsteady state conditions	Particle Material	Particle Shape	Particle Size (mm)
1943	Gamson et al. [1]	Evaporation of water	Steady	Celite	Sphere	2.3, 3.0, 5.6, 8.4 11.6
					Cylinder	4.1×4.8, 6.8×8. 9.8×11.7, 14.0×12.5, 18.8×16.9
1943	Hurt [2]	Evaporation of water	Steady		Cylinder	9.5×9.5
1945	Wilke and Hougen [3]	Evaporation of water	Steady	Celite	Cylinder	3.1×3.1, 4.8×4. 6.6×7.2, 9.7×8. 13.4×12.8, 15.1×16.3, 18.2×16.9
1952	Eichhorn and White [4]	High frequency dielectric heating particles	Steady	Dowex-50	Sphere	0.1, 0.3, 0.4, 0.5, 0.7
1954	Satterfield and Resnick [5]	Decomposition of H_2O_2	Steady	Polished catalytic metal	Sphere	5.1
1957	Galloway et al. [6]	Evaporation of water	Steady	Celite	Sphere	17.1
1958	Glaser and Thodos [7]	Heating metallic particles by	Steady	Monel	Sphere	4.8
		passing electric current through the beds		Brass Steel	Sphere Cylinder Cube	6.4, 7.9 6.4, 9.5 6.4, 9.5
1958	Baumeister and Bennett [8]	High frequency induction heating particles	Steady	Steel	Sphere	3.9, 6.3, 9.5
1960	De Acetis and Thodos [9]	Evaporation of water	Steady	Celite	Sphere	15.9
1961	Kunii and Smith [10]	Axial heat conduction in beds	Steady	Glass Sand	Sphere	0.1, 0.4, 0.6, 1.0 0.1, 0.2
1963	McConnachie and Thodos [11]	Evaporation of water	Steady	Celite	Sphere	15.9

TABLE 8.1 (*Continued*)

uid	Pr	Re	Determination of the heat transfer coefficients		Remarks
			Particle temperature	Fluid dispersion considered	
ir	0.72–0.75	100–4000	Surface assumed to be at wet-bulb temperature	No	–
ir	0.76	72–950	Measured	No	–
ir	0.73	45–250	Surface assumed to be at wet-bulb temperature	No	Heat transfer coefficients were not determined, but obtainable from their data
ir O_2	0.7 0.8	1–18	Measured	No	The measurements were criticized by Littman *et al.* [18]
apor mixture of $_2O_2$ and H_2O	1.0	15–160	Measured	No	–
ir	0.72	150–1200	Measured	No	–
ir $_2$ O_2	0.71 0.72 0.67	100–9200[b]	Measured	No	–
ir	0.7	200–10 400	Measured	No	The measurements were criticized by Jeffreson [28]
ir	0.72	32–2100	Measured	No	–
e, Air, O_2, Water		0.001–1	Continuous solid phase with axial heat conduction assumed	Yes	The anomalously low data were criticized by Littman *et al.* [18] and Gunn *et al.* [23]
ir	0.72	110–2500[b]	Measured	No	–

TABLE 8.1 (*Continued*)

Year	Investigator	Experimental method	Steady or unsteady state conditions	Particle		
				Material	Shape	Size (mm)
1963	Bradshaw and Myers [12]	Evaporation of water	Steady	Kaoline	Sphere	4.7
				AMT	Sphere	8.8
				Kaosorb	Cylinder	4.0×4.1
				Celite	Cylinder	4.2×4.2, 6.2×4.9
1963	Sen Gupta and Thodos [13]	Evaporation of water	Steady	Celite	Sphere	15.9
1964	Sen Gupta and Thodos [14]	Evaporation of water	Steady	Celite	Sphere	15.9
1967	Malling and Thodos [15]	Evaporation of water	Steady	Celite	Sphere	15.7–15.9
1967	Lindauer [16]	Frequency response	Unsteady	Steel	Sphere	1.0, 1.8, 3.2
				Tungsten	Sphere	0.5
1968	Handley and Heggs [17]	Step response	Unsteady	Steel	Sphere	3.2, 6.4, 9.5
					Cylinder	4.8×4.8, 6.4×6.4, 6.4×12.7
				Lead	Sphere	3.0, 6.1, 9.1
				Bronze	Sphere	9.5
				Soda glass	Sphere	6.1, 9.1
				Lead glass	Sphere	3.0
				Alumina–silica	Sphere	3.2
1968	Littman *et al.* [18]	Frequency response	Unsteady	Copper	Sphere	0.5, 0.6, 0.7, 1.1
				Glass	Sphere	0.5
				Lead	Sphere	2.0
1970	Bradshaw *et al.* [19]	Step response	Unsteady	Alumina	Sphere	13.2, 25.4
				Steel	Sphere	25.2
				Hematite	Sphere	11.1
1971	Goss and Turner [20]	Frequency response	Unsteady	Soda-lime glass	Sphere	4.0
				Borosilicate glass	Sphere	5.0
				Methyl meth-acrylate	Sphere	4.8

TABLE 8.1 (*Continued*)

luid	Pr	Re	Particle temperature	Fluid dispersion considered	Remarks
			Determination of the heat transfer coefficients		
air	0.7	400–6500[b]	Measured	No	–
air	0.72	800–2000	Measured	No	–
air	0.72	2000–6000	Measured	No	–
Air	0.71	185–8500	Measured	No	–
Air	0.7	23–18 200	No temperature gradient assumed in the particle	No	–
Air	0.7	80–4000	No temperature gradient assumed in the particle	No	The data were corrected by Jeffreson [28] for fluid dispersion
Air	0.7	2–100	Continuous solid phase with axial heat conduction assumed	Yes	The method of determining *Nu* data was criticized by Kaguei *et al.* [29]
Air, N₂	0.74	150–600[b]	Center-symmetric temperature profile assumed in the particle	Yes	–
Air	0.7	1600–3000	Center-symmetric temperature profile assumed in the particle	Yes	–

TABLE 8.1 (*Continued*)

Year	Investigator	Experimental method	Steady or unsteady state conditions	Particle		
				Material	Shape	Size (mm)
1973	Turner and Otten [21]	Frequency response	Unsteady	Soda-lime glass	Sphere	4.0
				Ceramic	Spheroid	7.4
				Sintered glass		4.6
				Fertilizer		3.1
				Iron ore		4.8
				Epoxy	Sphere	3.5
1974	Balakrishnan and Pei [22]	Microwave heating particles	Steady	Iron oxide	Sphere	6.4
				Nickel oxide	Sphere	6.4, 12.7
				Vanadium pentoxide	Sphere	4.8
					Cylinder	5.6×5.6, 5.6×8.?
				Nickel–molyb-denum oxide	Cylinder	3.2×6.4
				Cobalt–molyb-denum	Cylinder	3.2×6.4
1974	Gunn and De Souza [23]	Frequency response	Unsteady	Glass	Sphere	0.3, 0.5, 1.2, 2.2, 3.0, 6.0
				Steel	Sphere	3.2, 6.3
				Lead	Sphere	0.8
1975	Bhattacharyya and Pei [24]	Microwave heating particles	Steady	Ferric oxide	Sphere	3.2, 7.6
					Cylinder	5.1×5.1
1975	Cybulski et al. [25]	Radial heat conduction in beds	Steady	Silicon–copper	Irregular	0.1
1976	Wakao et al. [26]	Shot response	Unsteady	Polystyrene	Sphere	0.8
				Glass	Sphere	0.3, 1.1, 1.6
				Lead	Sphere	0.9
1981	Shen et al. [27]	Shot response	Unsteady	Glass	Sphere	1.3, 2.7

TABLE 8.1 (*Continued*)

Fluid	Pr	Re	Particle temperature	Fluid dispersion considered	Remarks
			Determination of the heat transfer coefficients		
Air	0.7	1200–4600	Center-symmetric temperature profile assumed in the particle	Yes	–
Air	0.7	340–4400[b]	Measured	No	–
Air	0.7	0.05–330	Center-symmetric temperature profile assumed in the particle	Yes	Nusselt numbers at $Re < 1$ were not determined
Air	0.7	110–830	Measured	No	–
Air	0.7	0.24–0.64	Continuous solid phase with radial heat conduction assumed	No	The anomalously low Nu data were criticized by Wakao *et al.* [30]
Air	0.7	0.2–6	Center-symmetric temperature profile assumed in the particle	Yes	No definite Nu data were obtained, but it was shown that Nu cannot be smaller than 0.1
Air	0.7	5.1–229	Center-symmetric temperature profile assumed in the particle	Yes	The results are shown in Figure 8.10

[a] Diluted beds, distended beds and data with a single particle layer are not included.
[b] $Re = D_p G/\mu$ except for Refs. [7] where $Re = S_p^{1/2} G/[\mu(1-\epsilon_b)\Psi]$, in which S_p is the particle surface area and Ψ is the shape factor; Refs. [11, 12, 22] where $Re = D_p G/[\mu(1-\epsilon_b)]$ and Ref. [19] where $Re = D_p G/[6\mu(1-\epsilon_b)]$.

A } Kunii and Smith	(1961)	
B		
C Cybulski *et al.*	(1975)	
D Eichhorn and White	(1952)	
E Littman *et al.*	(1968)	
F Gunn and De Souza	(1974)	
G Balakrishnan and Pei	(1974)	
H Bhattacharyya and Pei	(1975)	
I Glaser and Thodos	(1958)	
J Satterfield and Resnick	(1954)	

K Bradshaw *et al.*	(1970)	
L Handley and Heggs	(1968)	
(Hougen *et al.*	(1943, 45)	
{ Hurt	(1943)	
M { Galloway *et al.*	(1957)	
{ Thodos *et al.*	(1960–67)	
(Bradshaw and Myers	(1963)	
N Lindauer	(1967)	
O Baumeister and Bennett	(1958)	
P Turner *et al.*	(1971, 73)	

A: Liquid–Solid; B–P: Gas–Solid

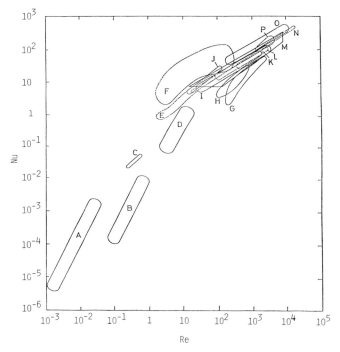

FIGURE 8.1   Heat transfer coefficients published in the literature.

8.1   A Review and Correction of the Data Obtained from Steady-state Measurements

Similar to the discussion in Chapter 4, the data selection is confined to the heat transfer measurements which satisfy the following conditions:

a) The particles in the bed are all active. Distended and diluted bed data not included.

b) The number of particle layers in a heat transfer bed are greater than two.

8.1.1 Simultaneous Heat and Mass Transfer Studies: Evaporation of Water and Diffusion-controlled Chemical Reactions at Particle Surfaces

Mass transfer data obtained from simultaneous heat and mass transfer studies have been used for the data correlation in Chapter 4. The corresponding heat transfer data which were determined without considering the effect of thermal dispersion are re-assessed in this section. The data include those obtained from measurements of the rates of evaporation of water by Gamson et al. [1], Hurt [2], Wilke and Hougen [3], Galloway et al. [6], Thodos et al. [9, 11, 13–15] and Bradshaw and Myers [12], as well as those determined from the catalytic decomposition of hydrogen peroxide on metal spheres by Satterfield and Resnick [5]. These measurements were made using solid particles with a constant surface temperature, T_{ps}, throughout the bed.

In packed beds with a constant particle surface temperature, the resistance to heat transfer resides only on the fluid side. If the contribution due to axial fluid thermal dispersion is neglected, the system may be described by the following heat balance equation under steady-state conditions (with the notation used in Chapter 6):

$$U \frac{dT_F}{dx} + \frac{h_p^{\dagger} a}{\epsilon_b C_F \rho_F} (T_F - T_{ps}) = 0 \qquad \text{for } 0 < x < L \qquad (8.1)$$

where $h_p^{\dagger}$ is the heat transfer coefficient with the axial fluid thermal dispersion coefficient, α_{ax}, equal to zero.

When a fluid, at a temperature T_{in} is flowing into a bed of length L, the exit temperature, T_{exit}, is

$$\frac{T_{ps} - T_{exit}}{T_{ps} - T_{in}} = \exp \left(-\frac{h_p^{\dagger} a L}{\epsilon_b C_F \rho_F U} \right). \qquad (8.2)$$

If, however, axial fluid thermal dispersion is taken into consideration, then, the heat balance equation becomes

$$U\frac{\mathrm{d}T_F}{\mathrm{d}x} + \frac{h_p a}{\epsilon_b C_F \rho_F}(T_F - T_{ps}) = \alpha_{ax}\frac{\mathrm{d}^2 T_F}{\mathrm{d}x^2}. \tag{8.3}$$

With the following Danckwerts boundary conditions:

$$U(T_F - T_{in}) = \alpha_{ax}\frac{\mathrm{d}T_F}{\mathrm{d}x} \qquad \text{at } x = 0 \text{ (inlet)} \tag{8.3a}$$

and

$$\frac{\mathrm{d}T_F}{\mathrm{d}x} = 0 \qquad \text{at } x = L \text{ (exit)} \tag{8.3b}$$

the exit fluid temperature is then

$$\frac{T_{ps} - T_{exit}}{T_{ps} - T_{in}} = \frac{4A \exp\left(\dfrac{LU}{2\alpha_{ax}}\right)}{(1+A)^2 \exp\left(A\dfrac{LU}{2\alpha_{ax}}\right) - (1-A)^2 \exp\left(-A\dfrac{LU}{2\alpha_{ax}}\right)} \tag{8.4}$$

where

$$A = \left(1 + \frac{4h_p a \alpha_{ax}}{\epsilon_b C_F \rho_F U^2}\right)^{1/2}. \tag{8.4a}$$

By applying Eqs. (8.2) and (8.4), the heat transfer coefficients, $h_p^\dagger$, reviewed in this section, can be converted into h_p, which takes into account the axial thermal dispersion effect. The values of the axial fluid thermal dispersion coefficient, α_{ax}, used in Eq. (8.4) are estimated from Eq. (6.28). All the published data, except those from articles by Gamson et al. [1], Hurt [2], and Bradshaw and Myers [12], in which no detailed information on bed height and/or void fraction is available, are thus corrected.

8.1.2   Temperature Measurements in Beds with no Heat Generating Particles

Kunii and Smith [10] and Cybulski et al. [25] determined heat transfer coefficients based on the C–S model from steady-state heat transfer measurements.

Heat transfer coefficients determined from axial heat transfer measurements by Kunii and Smith [10] were subjected to criticism by Littman *et al.* [18] and Gunn and De Souza [23]. They pointed out that the anomalously low heat transfer coefficients obtained by Kunii *et al.* were due to incorrect interpretation of an algebraic relationship between the heat transfer coefficient and the effective thermal conductivity of the bed. Gunn and De Souza [23] further elucidated that, if Kunii *et al.* had interpreted the algebraic relationship correctly, infinitely large heat transfer coefficients would have been obtained.

Cybulski *et al.* [25] obtained anomalously low heat transfer coefficients by fitting the steady-state radial gas temperatures measured at the bed exit and the corresponding gas temperatures estimated based on the C–S model. Wakao *et al.* [30] have shown that the gas and solid temperature profiles at the bed exit predicted based on the C–S model do not agree with each other at all. As a matter of fact, the solid and gas phase temperatures, under steady-state conditions and in a bed with no heat generating particles, are considered to be identical with each other, as illustrated in Figure 5.1. Thus, the C–S model cannot be applied to steady-state heat transfer measurements unless there is a heat source or sink existing in the solid particles. This is demonstrated in the example as follows.

Example 8.1

Suppose the steady-state radial heat transfer measurements are made with the following solid–gas system:

Fluid: air
Solid: glass beads of 0.5 mm diameter
$k_F = 0.030$ W m^{-1} K^{-1}
$k_S = 0.88$ W m^{-1} K^{-1}
$C_S = 670$ J kg^{-1} K^{-1}
$\rho_S = 2500$ kg m^{-3}
Wall temperature, $T_w = 100°$C
Fluid temperature at bed inlet, $T_0 = 50°$C
Radius of bed, $R_T = 2$ cm
Length of bed, $L = 2$ cm
Void fraction, $\epsilon_b = 0.4$
$Re = 0.1$.

Predict the radial temperature profile at the bed exit. Also, examine whether the C–S model may be applied to steady-state heat transfer in a packed bed. Assume no heat transfer resistance at the column wall (this is the same assumption made by Cybulski *et al.* [25]).

SOLUTION

(i) *Temperature at the bed exit*

Bed temperatures, under steady-state conditions, may be estimated from Eq. (5.1) based on the Single Phase model. At such a low flow rate ($Re = 0.1$) axial heat conduction should be taken into consideration. Using the same boundary conditions employed by Cybulski *et al.* [25]:

$$T = T_0 \qquad \text{at } x = 0 \, (\text{inlet})$$

$$\frac{\partial T}{\partial r} = 0 \qquad \text{at } r = 0$$

$$T = T_w \qquad \text{at } r = R_T$$

together with the Danckwerts condition at the bed exit:

$$\frac{\partial T}{\partial x} = 0 \qquad \text{at } x = L$$

the solution to Eq. (5.1) for a packed bed of finite length, L, gives

$$\frac{T_w - T_L}{T_w - T_0} = 2 \sum_{n=1}^{\infty} F_n \frac{J_0(a_n r / R_T)}{a_n J_1(a_n)} \tag{8.5}$$

where

$$F_n = \frac{\alpha_n - \beta_n}{\alpha_n \exp(-\beta_n) - \beta_n \exp(-\alpha_n)} \tag{8.5a}$$

$$\alpha_n = A(1 + B_n) \tag{8.5b}$$

$$\beta_n = A(1 - B_n) \tag{8.5c}$$

$$A = \frac{GC_F L}{2k_{eax}} \tag{8.5d}$$

$$B_n = \left[1 + 4 \left(\frac{a_n}{GC_FR_T} \right)^2 k_{er} k_{eax} \right]^{1/2}$$

(8.5e)

and a_n is an n-th root of

$$J_0(a_n) = 0.$$

(8.6)

The bed exit temperature, T_L, calculated from Eq. (8.5), is represented by the solid lines in Figures 8.2(a)–(d).

(ii) *Solid and gas phase temperatures from the C-S model*
Under steady-state conditions the fundamental equations based on the C-S model are:

$$GC_F \frac{\partial T_F}{\partial x} - k_{eF} \left[\frac{1}{r} \frac{\partial}{\partial r} \left(r \frac{\partial T_F}{\partial r} \right) + \gamma \frac{\partial^2 T_F}{\partial x^2} \right] + h_p a(T_F - T_S) = 0 \quad (8.7)$$

$$k_{eS} \left[\frac{1}{r} \frac{\partial}{\partial r} \left(r \frac{\partial T_S}{\partial r} \right) + \frac{\partial^2 T_S}{\partial x^2} \right] + h_p a(T_F - T_S) = 0.$$

(8.8)

At this low flow rate $(Re = 0.1)$, the effective radial and axial fluid phase thermal conductivities are considered to be the same, i.e. $\gamma = 1$. If axial fluid phase thermal conduction is ignored, then, γ is set equal to zero.

The C-S model was employed by Littman *et al.* [18] and Vortmeyer and Schaefer [43] in their studies of axial unsteady-state heat transfer in packed beds. Let us assume that their boundary conditions may be used for steady-state heat transfer as well. The axial boundary conditions of Cybulski *et al.* [25], Littman *et al.* [18], and Vortmeyer and Schaefer [43] are designated as Case A, C and D, respectively, and are shown in Table 8.2. Case B which has the boundary conditions somewhat similar to those of Case A, except that axial fluid phase heat conduction is considered, is also included in Table 8.2.

Note that, in solving the fundamental equations, Littman *et al.* considered the axial fluid phase conduction term, while Vortmeyer and Schaefer ignored it. Cybulski *et al.* also neglected the axial fluid phase heat conduction term in their application of the C-S model for radial steady-state heat transfer measurements.

In addition to the axial boundary conditions given in Table 8.2, the following radial conditions are needed for solving Eqs. (8.7) and (8.8):

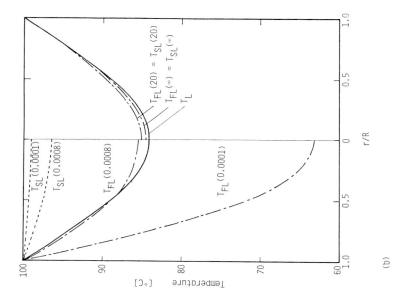

(b)

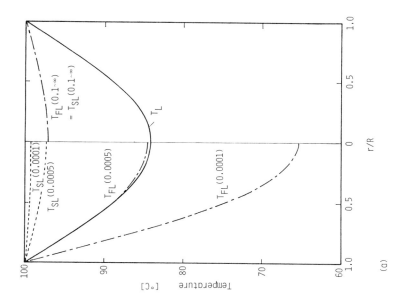

(a)

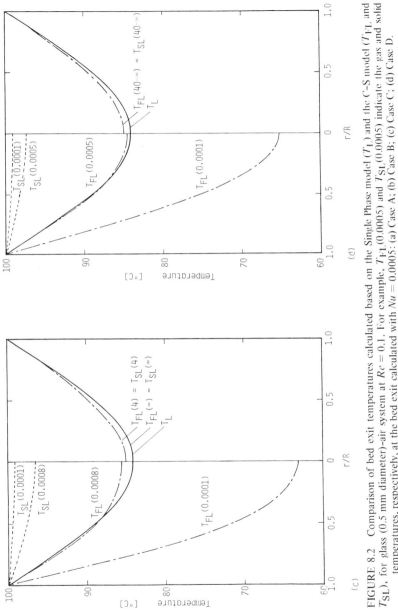

FIGURE 8.2 Comparison of bed exit temperatures calculated based on the Single Phase model (T_I) and the C–S model (T_{FI} and T_{SL}), for glass (0.5 mm diameter)–air system at $Re = 0.1$. For example, $T_{FL}(0.0005)$ and $T_{SL}(0.0005)$ indicate the gas and solid temperatures, respectively, at the bed exit calculated with $Nu = 0.0005$: (a) Case A; (b) Case B; (c) Case C; (d) Case D.

TABLE 8.2
Comparison of axial boundary conditions for C-S models.

Case	In fundamental Eqs. (8.7) and (8.8) γ	Axial boundary conditions				Conditions for $T_S(x=0)$, $T_F(x=L)$ and $T_S(x=L)$ are those assumed by
		x = 0		x = L		
		T_F	T_S	T_F	T_S	
A	0	T_0	$\dfrac{\partial T_S}{\partial x} = 0$	—	$\dfrac{\partial T_S}{\partial x} = 0$	Cybulski et al. [25] for steady-state analysis
B	1	T_0	$\dfrac{\partial T_S}{\partial x} = 0$	$\dfrac{\partial T_F}{\partial x} = 0$	$\dfrac{\partial T_S}{\partial x} = 0$	
C	1	T_0	$-k_{eS}\dfrac{\partial T_S}{\partial x} = (1-\epsilon_b)\,h_p(T_F - T_S)$	$\dfrac{\partial T_F}{\partial x} = 0$	$k_{eS}\dfrac{\partial T_S}{\partial x} = (1-\epsilon_b)\,h_p(T_F - T_S)$	Littman et al. [18] for unsteady-state analysis
D	0	T_0	$-k_{eS}\dfrac{\partial T_S}{\partial x} = (1-\epsilon_b)\,h_p(T_F - T_S)$	—	$k_{eS}\dfrac{\partial T_S}{\partial x} = (1-\epsilon_b)\,h_p\beta(T_F - T_S)$ $\beta = \dfrac{1}{1+(1-\epsilon_b)\,h_p/(GC_F)}$	Vortmeyer and Schaefer [43] for unsteady-state analysis

$$\frac{\partial T_F}{\partial r} = \frac{\partial T_S}{\partial r} = 0 \qquad \text{at } r = 0$$

$$T_F = T_S = T_w \qquad \text{at } r = R_T.$$

The temperatures at the bed exit, T_{FL} and T_{SL} are, hence,

$$\frac{T_w - T_{FL}}{T_w - T_0} = 2 \sum_{n=1}^{\infty} \frac{J_0(a_n r/R_T)}{a_n J_1(a_n)} \sum_{i=1}^{3+\gamma} P_{ni} \exp (m_i) \qquad (8.9)$$

$$\frac{T_w - T_{SL}}{T_w - T_0} = 2 \sum_{n=1}^{\infty} \frac{J_0(a_n r/R_T)}{a_n J_1(a_n)} \sum_{i=1}^{3+\gamma} Q_{ni} \exp (m_i) \qquad (8.10)$$

where a_n is a root of Eq. (8.6), and m_i is a root of the following equation:

$$\gamma m_i^4 - \frac{GC_F L}{k_{eF}} m_i^3 - \left[(1+\gamma) \left(\frac{L}{R} \right)^2 a_n^2 + \left(\frac{1}{k_{eF}} + \frac{\gamma}{k_{eS}} \right) h_p a L^2 \right] m_i^2$$

$$+ \frac{GC_F L}{k_{eF}} \left[\left(\frac{L}{R} \right)^2 a_n^2 + \frac{h_p a L^2}{k_{eS}} \right] m_i + \left(\frac{L}{R} \right)^4 a_n^4$$

$$+ \left(\frac{1}{k_{eF}} + \frac{1}{k_{eS}} \right) \frac{h_p a L^4}{R^2} a_n^2 = 0. \qquad (8.11)$$

P_{ni} and Q_{ni} are obtained from

$$Q_{ni} = -\frac{k_{eF}}{h_p a L^2} \left\{ \gamma m_i^2 - \frac{GC_F L}{k_{eF}} m_i - \left[\left(\frac{L}{R} \right)^2 a_n^2 + \frac{h_p a L^2}{k_{eF}} \right] \right\} P_{ni}$$

$$\sum_{i=1}^{3+\gamma} P_{ni} = 1$$

$$\sum_{i=1}^{3+\gamma} \left[\frac{L h_p \lambda}{k_{eS}} P_{ni} + \left(m_i - \frac{L h_p \lambda}{k_{eS}} \right) Q_{ni} \right] = 0$$

$$\sum_{i=1}^{3+\gamma} \left[\frac{L h_p \lambda'}{k_{eS}} P_{ni} - \left(m_i + \frac{L h_p \lambda'}{k_{eS}} \right) Q_{ni} \right] \exp (m_i) = 0$$

$$\left. \right\} (8.12)$$

and

$$\sum_{i=1}^{3+\gamma} m_i \exp(m_i) P_{ni} = 0 \tag{8.13}$$

where γ, λ and λ' for Cases A–D are tabulated as follows:

Case	γ	λ	λ'
A	0	0	0
B	1	0	0
C	1	$1 - \epsilon_b$	$1 - \epsilon_b$
D	0	$1 - \epsilon_b$	$(1 - \epsilon_b)\beta$

See Table 8.2 for β.

In the case of $\gamma = 0$, P_{ni} and Q_{ni} can only be determined from the relationships given by Eq. (8.12).

The effective fluid phase thermal conductivity, k_{eF}, is given by Eq. (7.27). The effective solid phase thermal conductivity, k_{eS}, is assumed to be related to the effective axial thermal conductivity, k_{eax}, and k_{eF} by:

$$k_{eax} = k_{eF} + k_{eS}. \tag{8.14}$$

At $Re = 0.1$, $k_{eax} \simeq k_e^o$, where k_e^o is the effective thermal conductivity of a quiescent bed. Hence,

$$k_{eS} = k_e^o - k_{eF}. \tag{8.15}$$

The solid temperature (T_{SL}) and gas temperature (T_{FL}) profiles at the bed exit are then computed. In Figures 8.2(a)–(d), the predicted profiles are compared with the temperature (T_L) profile estimated in Solution (i) based on the Single Phase model.

In Cases B–D, it is found that

a) T_L is in good agreement with T_{FL} predicted with certain small Nusselt numbers, but T_{FL} is different from the corresponding T_{SL} value.

b) T_L concurs well with both T_{FL} and T_{SL} computed with infinitely large Nusselt numbers.

Therefore, if a measured temperature profile is compared with the computed T_{FL} values, as done by Cybulski et al. [25], it is expected that two different Nusselt values, one very small and the other infinitely large, will be found. The C–S model reduces to the Single Phase model when $T_S \simeq T_F$ or the particle-to-fluid heat transfer coefficient, h_p, is infinitely large. Since the packed bed temperature under steady-state conditions is described by the Single Phase model, the application of the C–S model to steady-state heat transfer is expected to yield, superficially, infinite values of Nusselt number.

However, in Case A of Cybulski et al. [25], infinitely large Nusselt numbers are not obtained; T_{FL} and T_{SL} estimated with large Nusselt numbers are considerably different from T_L. This is due to the assumptions inherent in Case A: axial fluid phase conduction is neglected, and the solid phase is assumed to be adiabatic at both ends. From a comparison of T_L and T_{FL}, small Nusselt numbers are obtained. The Nusselt numbers determined by Cybulski et al. are of this magnitude.

However, attention should be paid to the fact that in Cases A–D, when Nusselt numbers are small, the T_{SL} value is entirely different from the corresponding T_{FL} value. As mentioned already, T_{FL} and T_{SL} should be the same under steady-state conditions. The small Nusselt numbers, obtained from the fitting of T_L and T_{FL}, do not represent the intrinsic heat transfer coefficients.

Therefore, it is concluded that the C–S model cannot be applied to steady-state heat transfer analysis. For reference, the Nusselt numbers erroneously obtained from the fitting of T_L and T_{FL} are shown in Figure 8.3.   (End of Example)

8.1.3   Heat Transfer Between Heat Generating Particles and Fluid

In their heat transfer measurements, Eichhorn and White [4], Baumeister and Bennett [8] and Pei et al. [22, 24] employed high frequency heating to generate heat in solid particles.

According to Eichhorn and White, solid temperature profiles in the bed are a linear function of the axial distance; thus, the solid temperature at the bed exit can be estimated by linear extrapolation. They assumed that the temperature difference between the solid and gas phases throughout the bed was the same as the difference between the extrapolated solid temperature and the measured gas temperature at the bed exit. With these assumptions they determined the heat transfer coefficients. Their results

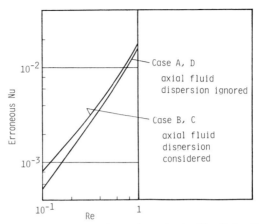

FIGURE 8.3   Erroneous Nusselt numbers obtained by fitting T_L and T_{FL}.

indicate that the temperature difference between the two phases is relatively small (0.5 to 2.1°C). Considering the errors (±0.6°C) associated with the measured solid temperature profiles, the heat transfer coefficients determined are, therefore, not reliable enough. Moreover, it is doubtful whether the solid temperature at the bed exit is predicted by simple extrapolation. To verify this, the temperature profile in the bed is analysed based on the D–C model as follows:

When heat is generated at a constant rate in particles, the steady-state fluid and solid temperatures in a bed of finite length are described by Eqs. (6.22) and (6.23) according to the D–C model. With the Danckwerts boundary conditions, Eqs. (8.3a) and (8.3b), the solutions to Eqs. (6.22) and (6.23) are expressed as:

$$T_F = T_{in} + \frac{L}{U}\left(\frac{q_v}{C_F\rho_F}\right)\frac{aR}{3\epsilon_b}$$

$$\times\left(\frac{x}{L} + \frac{\alpha_{ax}}{LU}\left\{1 - \exp\left[-\frac{LU}{\alpha_{ax}}\left(1 - \frac{x}{L}\right)\right]\right\}\right) \qquad (8.16)$$

and

$$T_S = T_F + \frac{q_vR}{3h_p} + \frac{q_v}{6k_S}(R^2 - r^2). \qquad (8.17)$$

where q_v is the rate of heat generation per unit volume of solid. From Eq. (8.17), the average (volume mean) particle temperature is

$$\bar{T}_S = T_F + \frac{q_v R}{3h_p} \left(1 + \frac{Rh_p}{5k_S}\right). \tag{8.18}$$

Equation (8.16) shows that the fluid temperature increases almost linearly in the axial direction but levels off before it reaches the bed exit. Accordingly, Eq. (8.18) indicates that the temperature gradient of $\bar{T}_S$ should be approximately constant within the bed and becomes zero at the bed exit. The bed exit effect, however, was ignored by Eichhorn and White [4] in their extrapolation of the solid temperatures to the bed exit value. Hence, the temperature difference between the gas and solid predicted by Eichhorn and White is expected to be larger than the actual difference in the bed.

In their investigation Baumeister and Bennett [8] found significant temperature gradients in both the radial and axial directions in the bed. Their results on the large radial temperature differences were subjected to criticism by Jeffreson [28]. Pei *et al.* [22, 24], on the other hand, found that the solid temperature was uniform throughout the bed; based on this, they evaluated the heat transfer coefficients. Their observed uniform solid temperature in the bed contradicts the findings of previous investigators, e.g. Eichhorn and White [4], and Baumeister and Bennett [8].

Glaser and Thodos [7] applied an electric current directly through a bed of metal spheres to generate heat in the particles. However, in applying direct electric heating there is a chance that heat transfer may occur only at, or near, the solid–solid contact points where most of the heat is generated. If this is the case, then the measured heat transfer coefficients would be different from those we are concerned with.

Therefore, based on the above considerations, the data reviewed in this section are not to be included in the data correlation.

8.2   A Review and Correction of the Data Obtained from Unsteady-state Measurements

Heat transfer coefficients in unsteady-state packed bed systems are usually obtained using step, frequency and shot response techniques. The determination of transfer coefficients from the response measurements is based on the unsteady-state models: the Schumann model, the C–S model and the D–C model, discussed in Chapter 7.

The Schumann model [44], which assumes no temperature gradient in solid particles and no dispersion in fluid phases, is not realistic and is inadequate for describing unsteady-state heat transfer. Both the C–S model and the original D–C model (with α'_{ax} from Eq. 6.3) have been shown to be incapable of predicting accurate transfer coefficients at low Reynolds numbers. So far, the modified D–C model (with α_{ax} from Eq. 6.28) is the most successful model proposed.

Owing to the varying assumptions underlying the different models, the values of the heat transfer coefficients based on these models are generally incompatible, especially at low Reynolds numbers; thus, they cannot be compared simply on an equal basis.

Therefore, the published unsteady-state heat transfer data are treated according to the modified D–C model and then correlated together with the re-assessed steady-state heat transfer data.

8.2.1 Heat Transfer Data obtained from Step Response Measurements

Step response measurements were made by Handley and Heggs [17], Furnas [45], Saunders and Ford [46], Löf and Hawley [47], and Coppage and London [48], in the determination of heat transfer coefficients. They all determined heat transfer coefficients based on the Schumann model. Bradshaw *et al.* [19] also determined heat transfer coefficients based on the Schumann model, and then converted the data into those according to the original D–C model. In the earlier studies [45–48], the heat transfer coefficients determined by graphical methods are not reliable. Therefore, only the data of the two relatively recent measurements by Handley and Heggs [17], and Bradshaw *et al.* [19] are considered in the modified D–C model. The data of Handley and Heggs have already been re-evaluated in Section 7.5.

The heat transfer coefficient data of Bradshaw *et al.* [19] are revised as follows: Assuming that a step temperature change is imposed on the fluid entering the packed bed, the response signals are predicted with the data presented in their paper. The predicted signals are denoted by $(T_F^{II})_{Schumann}$. The response signals based on the modified D–C model, $(T_F^{II})_{D-C}$, are also computed with the α_{ax} values from Eq. (6.28) and various assumed Nusselt values. By fitting $(T_F^{II})_{Schumann}$ and $(T_F^{II})_{D-C}$ in the time domain, the correct Nusselt values of the modified D–C model,

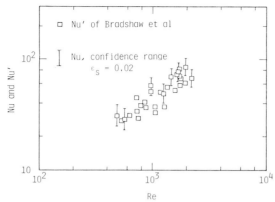

FIGURE 8.4   *Nu'* of Bradshaw *et al.* [19] and *Nu* re-evaluated based on the modified D–C model.

Nu, are determined. The re-evaluated values of *Nu* are plotted together with their corresponding data, *Nu'*, in Figure 8.4. For some of the data, the confidence ranges are indicated in terms of the root-mean-square-error, ϵ_s, equal to 0.02, as defined by Eq. (7.30). As illustrated, the agreement between *Nu* and *Nu'* is fairly good. This is not unexpected since the measurements were made at high Reynolds numbers (*Re* = 490 to 2200), thereby, Eqs. (6.3) and (6.28) give almost the same axial fluid thermal dispersion coefficient values.

8.2.2   Heat Transfer Data obtained from Frequency Response Measurements

Heat transfer coefficients were obtained from frequency response measurements by Lindauer [16], Littman *et al.* [18], Goss and Turner [20], Turner and Otten [21], Gunn and De Souza [23], and Littman and Sliva [49].

Lindauer [16] employed the fundamental equations based on the Schumann model to determine heat transfer coefficients, but no information on the frequency range is given in his paper. Therefore, the heat transfer data cannot be converted into those based on the modified D–C model. The data of Littman *et al.* [18], obtained on the C–S model, are examined and revised in Section 7.4, according to the modified D–C model. Goss and Turner [20], Turner and Otten [21], and Gunn and De Souza [23], all reported heat transfer coefficients and axial fluid thermal dispersion coefficients based on the D–C model. The following tests are carried out to assess the reliability of these data:

8.2.2.1   *Confidence test on the data of Gunn and De Souza* [23]

Using the reported α_{ax} and Nu data of Gunn and De Souza [23], frequency response signals, $(T_F^{II})_{Ref}$, are first predicted from Eqs. (6.10) and (7.19). For the same system, the response signals, $(T_F^{II})_{D-C}$, are also computed from Eqs. (6.10) and (7.19) with various assumed values of α_{ax} and Nu. The signals, $(T_F^{II})_{D-C}$ and $(T_F^{II})_{Ref}$, are then compared in terms of the following root-mean-square-error:

$$\epsilon_f^* = \left\{ \frac{\displaystyle\int_0^{2\pi/\omega} [(T_F^{II})_{Ref} - (T_F^{II})_{D-C}]^2 \, dt}{\displaystyle\int_0^{2\pi/\omega} [(T_F^{II})_{Ref}]^2 \, dt} \right\}^{1/2}. \tag{8.19}$$

The resulting error maps are constructed as shown in Figures 8.5(a) and (b). In Figure 8.5(a) ($Re \simeq 11$), the horizontal contour with $\epsilon_f^* = 0.05$ indicates that the precise value of Nusselt number cannot be determined. Although Gunn and De Souza obtained $Nu \simeq 50$, the contour shows that the Nusselt number can be any value greater than about 10. In Figure 8.5(b) ($Re \simeq 33$), the contour with $\epsilon_f^* = 0.05$ reveals that the Nusselt value should fall in the range 20 to 90. The confidence ranges of Nusselt numbers, thus estimated, are shown in Figure 8.6. As illustrated, the ranges are large at low Reynolds numbers; moreover, their original data show considerable scattering. Therefore, their data will not be considered in the correlation in Section 8.3.

8.2.2.2   *Confidence test on the data of Turner* et al. [20, 21]

No detailed information on the frequencies employed in the measurements is given in the papers of Goss and Turner [20], and Turner and Otten [21]. There is, however, a simulated example presented in the paper of Goss and Turner. This example ($Re \simeq 950$) is subjected, therefore, to the sensitivity test. As shown in Figure 8.7, the contour with an error, $\epsilon_f^* = 0.05$, is steep. As far as this example is concerned, the Nusselt number determined is considered to be reliable enough. In fact, the measurements made by Turner *et al.* are at Reynolds numbers as high as 1200 to 4600. At such

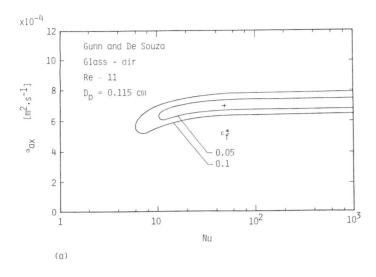

(a)

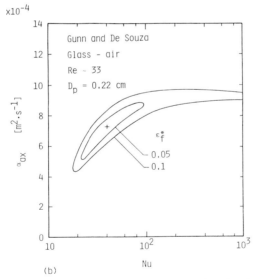

(b)

FIGURE 8.5 Error maps in the plot of α_{ax} versus Nu, for data of Gunn and De Souza [23] (+ shows the data obtained by them): (a) $Re \simeq 11$, $\epsilon_b = 0.4$, $D_p = 1.15$ mm, $L = 3$ cm and the amplitude ratio $= 0.3$; (b) $Re \simeq 33$, $\epsilon_b = 0.4$, $D_p = 2.2$ mm, $L = 3$ cm and the amplitude ratio $= 0.3$.

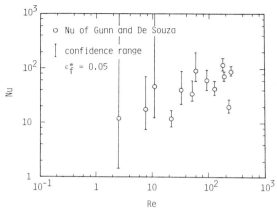

FIGURE 8.6 *Nu* data of Gunn and De Souza [23] with the confidence range indicating $\epsilon_f^* = 0.05$.

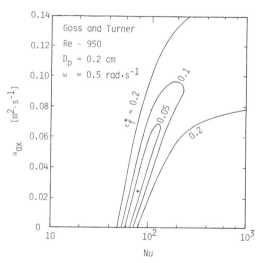

FIGURE 8.7 Error map for a simulated example of Goss and Turner [20] (+ shows the data obtained by them); $Re \simeq 950$, $\omega = 0.5$ rad s^{-1}, other data listed in their Table 1.

high flow rates, the effect of axial fluid thermal dispersion on the overall heat transfer is small. Hence, heat transfer coefficients obtained under such conditions are usually reliable and quite consistent. The data of Turner *et al.* will be included in the data correlation.

8.2.3 Heat Transfer Data obtained from Shot Response Measurements

From the analysis of shot response measurements, Wakao *et al.* [26, 41] examined heat transfer coefficients based on the modified D–C model with α_{ax} from Eq. (6.28). No definite Nusselt numbers could be obtained, but it was found that they fall within the range 0.1 to ∞ over the Reynolds number range 0.2 to 6.

Based on the modified D–C model, Shen *et al.* [27] determined heat transfer parameters from curve fitting in the time domain using the one-shot input technique. Their results are shown in Figure 8.8 for $Re = 5.1$. As depicted, the axial fluid thermal dispersion coefficient, α_{ax}, and the particle-to-fluid heat transfer coefficient, h_p, cannot be determined simultaneously from a single measurement. It is also revealed that α_{ax} is almost independent of the Nusselt number at a Nusselt number greater than about 3, according to the contour with a root-mean-square-error, ϵ, defined by Eq. (6.18), of 0.03.

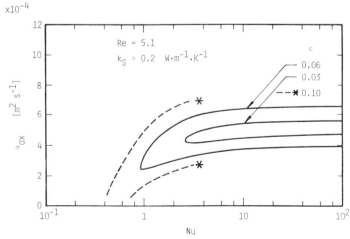

FIGURE 8.8 Error map in the plot of α_{ax} versus Nu, from Shen *et al.* [27]; $Re = 5.1$, $\epsilon_b = 0.39$, $D_p = 1.3$ mm (glass beads) and $L = 1.3$ cm.

8.3 Correlation of Nusselt Numbers

Figure 8.1 reveals a mixture of heat transfer data obtained from the different models. As mentioned already, some of the data are less reliable and some have been criticized for the improper methods employed in the analysis; these data will not be considered in the correlation. From the review given in the preceding sections, the reported heat transfer measurements which have satisfied the predetermined criteria are as follows:

a)   Steady-state measurements [3, 5, 6, 9, 11, 13-15];

b)   Unsteady-state measurements [17, 19-21].

The heat transfer coefficients re-evaluated according to the modified D–C model are plotted in Figure 8.9. As depicted, the re-evaluated particle-to-fluid heat transfer coefficients, expressed in terms of Nusselt numbers, are quite consistent and compatible. The values are considerably higher than their corresponding original values, in particular, at low Reynolds number. More importantly, the recalculated values show no tendency to decrease further with decreasing Reynolds number at low Reynolds number, and as the trend predicts, a limiting Nusselt number is approached at zero flow rate.

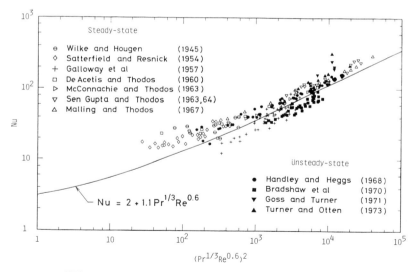

FIGURE 8.9   Correlation of re-evaluated Nusselt numbers.

Based on an analogy with Eq. (4.11) for mass transfer, Wakao *et al.* [50] proposed the following correlation (solid line in Figure 8.9):

$$Nu = 2 + 1.1 Pr^{1/3} Re^{0.6}. \qquad (8.20)$$

At lower Reynolds numbers, Figure 8.9 shows that the fitting is not as good as in the case of the mass transfer coefficient expressed in terms of Sherwood number, shown in Figure 4.4. This is not unusual in view of the fact that heat transfer measurements and determination of heat transfer cofficients are often more difficult than mass transfer measurements. For instance, the ratio, J_{Heat}/J_{Mass}, has been found to be 1.37 and 1.51 by Satterfield and Resnick [5] and De Acetis and Thodos [9], respectively; on the other hand, McConnachie and Thodos [11], Sen Gupta and Thodos [13, 14] and Malling and Thodos [15] found the ratio to be approximately 1.0. Considering all these, we may say that the heat transfer data shown in Figure 8.9 are well represented by Eq. (8.20).

The question of the limiting Nusselt number at zero flow rate has been the subject of much controversy. Different limiting Nusselt numbers have been estimated based on different models. Gunn and De Souza [23] obtained a limiting Nusselt value of 10 from frequency response measurement. But, Wakao *et al.* [26, 27, 41] have demonstrated, from one-shot measurements, and using curve fitting in the time domain, that no definite Nusselt values can be obtained at low Reynolds numbers. Figure 8.8 shows that, at $Re = 5.1$ and with $\epsilon = 0.03$, the Nusselt number varies from about 3 to ∞. The fact that any Nu value within this range will yield approximately the same value of α_{ax}, suggests that, at this low Reynolds number, particle-to-fluid heat transfer makes little contribution to the overall heat transfer in the system. This is further demonstrated by the Nu–Re relationship, given in Figure 8.10 obtained by Shen *et al.* [27]. It appears that the limiting Nusselt value may be somewhat higher than that predicted according to Eq. (8.20). However, as indicated, the confidence range increases significantly at lower Reynolds number. The high uncertainty in Nusselt values, at low flow rates, again implies the insignificant role of particle-to-fluid heat transfer in the overall heat transfer process. This deduction is not unreasonable considering the fact that, at low flow rates, a particle and its surrounding envelope of fluid are likely to be in thermal equilibrium. The authors feel that although there should be a limit to the decrease in Nusselt number with lowering Reynolds number, the particular limiting value is not practically important. For this reason, the relationship

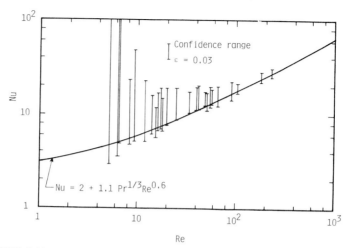

FIGURE 8.10   Confidence range in the *Nu-Re* relationship, from Shen *et al.* [27].

according to Eq. (8.20), which is based on a heat–mass analogy and predicts a limiting Nusselt number of two, is recommended.

REFERENCES

[1]   B. W. Gamson, G. Thodos and O. A. Hougen, *Trans. Amer. Inst. Chem. Eng.* **39**, 1 (1943).
[2]   D. M. Hurt, *Ind. Eng. Chem.* **35**, 522 (1943).
[3]   C. R. Wilke and O. A. Hougen, *Trans. Amer. Inst. Chem. Eng.* **41**, 445 (1945).
[4]   J. Eichhorn and R. R. White, *Chem. Eng. Prog. Symp. Ser.* **48** (No. 4), 11 (1952).
[5]   C. N. Satterfield and H. Resnick, *Chem. Eng. Prog.* **50**, 504 (1954).
[6]   L. R. Galloway, W. Komarnicky and N. Epstein, *Can. J. Chem. Eng.* **35**, 139 (1957).
[7]   M. B. Glaser and G. Thodos, *AIChE J.* **4**, 63 (1958).
[8]   E. B. Baumeister and C. O. Bennett, *AIChE J.* **4**, 69 (1958).
[9]   J. De Acetis and G. Thodos, *Ind. Eng. Chem.* **52**, 1003 (1960).
[10]   D. Kunii and J. M. Smith, *AIChE J.* **7**, 29 (1961).
[11]   J. T. L. McConnachie and G. Thodos, *AIChE J.* **9**, 60 (1963).
[12]   R. D. Bradshaw and J. E. Myers, *AIChE J.* **9**, 590 (1963).
[13]   A. Sen Gupta and G. Thodos, *AIChE J.* **9**, 751 (1963).
[14]   A. Sen Gupta and G. Thodos, *Ind. Eng. Chem. Fund.* **3**, 218 (1964).
[15]   G. F. Malling and G. Thodos, *Int. J. Heat Mass Transfer* **10**, 489 (1967).
[16]   G. C. Lindauer, *AIChE J.* **13**, 1181 (1967).
[17]   D. Handley and P. J. Heggs, *Trans. Inst. Chem. Eng.* **46**, T251 (1968).
[18]   H. Littman, R. G. Barile and A. H. Pulsifer, *Ind. Eng. Chem. Fund.* **7**, 554 (1968).

[19] A. V. Bradshaw, A. Johnson, N. H. McLachlan and Y. T. Chiu, *Trans. Inst. Chem. Eng.* **48**, T77 (1970).
[20] M. J. Goss and G. A. Turner, *AIChE J.* **17**, 590 (1971).
[21] G. A. Turner and L. Otten, *Ind. Eng. Chem. Process Des. Dev.* **12**, 417 (1973).
[22] A. R. Balakrishnan and D. C. T. Pei, *Ind. Eng. Chem. Process Des. Dev.* **13**, 441 (1974).
[23] D. J. Gunn and J. F. C. De Souza, *Chem. Eng. Sci.* **29**, 1363 (1974).
[24] D. Bhattacharyya and D. C. T. Pei, *Chem. Eng. Sci.* **30**, 293 (1975).
[25] A. Cybulski, M. J. Van Dalen, J. W. Verkerk and P. J. Van Den Berg, *Chem. Eng. Sci.* **30**, 1015 (1975).
[26] N. Wakao, S. Tanisho and B. Shiozawa, *Kagaku Kogaku Ronbunshu* **2**, 422 (1976).
[27] J. Shen, S. Kaguei and N. Wakao, *Chem. Eng. Sci.* **36**, 1283 (1981).
[28] C. P. Jeffreson, *AIChE J.* **18**, 409 (1972).
[29] S. Kaguei, B. Shiozawa and N. Wakao, *Chem. Eng. Sci.* **32**, 507 (1977).
[30] N. Wakao, S. Kaguei and H. Nagai, *Chem. Eng. Sci.* **32**, 1261 (1977).
[31] A. R. H. Cornish, *Trans. Inst. Chem. Eng.* **43**, T332 (1965).
[32] D. Kunii and M. Suzuki, *Int. J. Heat Mass Transfer* **10**, 845 (1967).
[33] P. A. Nelson and T. R. Galloway, *Chem. Eng. Sci.* **30**, 1 (1975).
[34] E. U. Schlünder, *Chem. Eng. Sci.* **32**, 845 (1977).
[35] H. Martin, *Chem. Eng. Sci.* **33**, 913 (1978).
[36] R. Pfeffer and J. Happel, *AIChE J.* **10**, 605 (1964).
[37] T. Miyauchi, *J. Chem. Eng. Japan* **4**, 238 (1971).
[38] J. P. Sørensen and W. E. Stewart, *Chem. Eng. Sci.* **29**, 827 (1974).
[39] E. U. Schlünder, Einführung in die Wärme- und Stoffübertragung, 2 Aufl. Vieweg-Verlag Braunschweig, p. 75 (1975).
[40] D. J. Gunn, *Int. J. Heat Mass Transfer* **21**, 467 (1978).
[41] N. Wakao, *Chem. Eng. Sci.* **31**, 1115 (1976).
[42] N. Wakao, S. Kaguei and B. Shiozawa, *Chem. Eng. Sci.* **32**, 451 (1977).
[43] D. Vortmeyer and R. J. Schaefer, *Chem. Eng. Sci.* **29**, 485 (1974).
[44] T. E. W. Schumann, *J. Franklin Inst.* **208**, 405 (1929).
[45] C. C. Furnas, *Ind. Eng. Chem.* **22**, 721 (1930).
[46] O. A. Saunders and H. Ford, *J. Iron Steel Inst.* **141**, 291 (1940).
[47] G. O. G. Löf and R. W. Hawley, *Ind. Eng. Chem.* **40**, 1061 (1948).
[48] J. E. Coppage and A. L. London, *Chem. Eng. Prog.* **52**, 57-F (1956).
[49] H. Littman and D. E. Sliva, *Proc. Int. Heat Transfer Conf.* Versailles, Vol. 7, CT1.4 (1970).
[50] N. Wakao, S. Kaguei and T. Funazkri, *Chem. Eng. Sci.* **34**, 325 (1979).

Appendix A. Physical Properties

Sources of the data: Chemical Engineers Handbook, 4 edn., Maruzen, Tokyo (1978).

A.1 Some Fundamental Physical Constants in SI Units

Quantity	Value in SI units[a]	Remarks
Avogadro number	$N_A = 6.022\ 045(31)$ $\times 10^{23}\ \text{mol}^{-1}$	
Boltzmann constant	$k\ \ = 1.380\ 622(44)$ $\times 10^{-23}\ \text{J K}^{-1}$	$k = R_g/N_A$
Gas constant	$R_g = 8.314\ 41(26)$ $\text{J K}^{-1}\ \text{mol}^{-1}$	$= 1.987\ 19\ \text{cal K}^{-1}\ \text{mol}^{-1}$ $= 6.236\ 32 \times 10^4\ \text{cm}^3$ $\text{mmHg K}^{-1}\ \text{mol}^{-1}$ $= 82.056\ 8\ \text{cm}^3\ \text{atm}$ $\text{K}^{-1}\ \text{mol}^{-1}$ $= 10.731\ 4\ \text{ft}^3\ \text{lb in}^{-2}$ $°\text{F}^{-1}\ \text{lb-mol}^{-1}$
Planck's constant	$h\ \ = 6.626\ 176(36)$ $\times 10^{-34}\ \text{J s}$	
Standard volume of ideal gas	$V_0 = 22.413\ 83(70)$ $\times 10^{-3}\ \text{m}^3\ \text{mol}^{-1}$	
Stefan–Boltzmann constant	$\sigma\ = 5.670\ 32(71)$ $\times 10^{-8}\ \text{W m}^{-2}\ \text{K}^{-4}$	$\sigma = 2\pi^5 k^4/(15\ h^3\ c^2)$
Velocity of light in a vacuum	$c\ \ = 2.997\ 924\ 58(1)$ $\times 10^8\ \text{m s}^{-1}$	

[a] The numbers in parentheses are the uncertainties in the last digits of the quoted value.

SI prefixes

	prefix				prefix
10^{-1}	deci	d	10	deca	da
10^{-2}	centi	c	10^2	hecto	h
10^{-3}	milli	m	10^3	kilo	k
10^{-6}	micro	μ	10^6	mega	M
10^{-9}	nano	n	10^9	giga	G
10^{-12}	pico	p	10^{12}	tera	T
10^{-15}	femto	f	10^{15}	peta	P
10^{-18}	atto	a	10^{18}	exa	E

A.2   Conversion Factors

SI units are shown in the first column. The digit is on FORTRAN E-format (for example, $E + 2 = 10^2$).

1)   Length (L)

m	cm	in	ft	yd
1	$1.000\,00\,E+2$	$3.937\,01\,E+1$	$3.280\,84\,E+0$	$1.093\,61\,E+0$
$1.000\,00\,E-2$	1	$3.937\,01\,E-1$	$3.280\,84\,E-2$	$1.093\,61\,E-2$
$2.540\,00\,E-2$	$2.540\,00\,E+0$	1	$8.333\,33\,E-2$	$2.777\,78\,E-2$
$3.048\,00\,E-1$	$3.048\,00\,E+1$	$1.200\,00\,E+1$	1	$3.333\,33\,E-1$
$9.144\,00\,E-1$	$9.144\,00\,E+1$	$3.600\,00\,E+1$	$3.000\,00\,E+0$	1

$1\,\text{Å} = 10^{-8}$ cm, $1\,\mu$ (micron) $= 10^{-3}$ mm $= 10^{-4}$ cm, 1 mile $= 5280$ ft $= 1609.3$ m.

2)   Mass (M)

kg	g	oz	lb
1	$1.000\,00\,E+3$	$3.527\,40\,E+1$	$2.204\,62\,E+0$
$1.000\,00\,E-3$	1	$3.527\,40\,E-2$	$2.204\,62\,E-3$
$2.834\,95\,E-2$	$2.834\,95\,E+1$	1	$6.250\,00\,E-2$
$4.535\,92\,E-1$	$4.535\,92\,E+2$	$1.600\,00\,E+1$	1

1 tonne (metric) $= 0.9842$ long ton (British) $= 1.102$ short ton (USA); 1 long ton (British) $= 2240$ lb $= 1.016\,05$ tonne (metric); 1 short ton (USA) $= 2000$ lb $= 0.907\,18$ tonne (metric).

3)   Specific volume $(L^3 M^{-1})$

$m^3\,kg^{-1}$	$cm^3\,g^{-1}$	$l\,kg^{-1}$	$in^3\,lb^{-1}$	$ft^3\,lb^{-1}$
1	$1.000\,00\,E+3$	$1.000\,00\,E+3$	$2.767\,99\,E+4$	$1.601\,85\,E+1$
$1.000\,00\,E-3$	1	1	$2.767\,99\,E+1$	$1.601\,85\,E-2$
$3.612\,73\,E-5$	$3.612\,73\,E-2$	$3.612\,73\,E-2$	1	$5.787\,04\,E-4$
$6.242\,80\,E-2$	$6.242\,80\,E+1$	$6.242\,80\,E+1$	$1.728\,00\,E+3$	1

4)   Density (ML^{-3})

$kg\,m^{-3}$	$g\,cm^{-3}$	$kg\,l^{-1}$	$lb\,in^{-3}$	$lb\,ft^{-3}$
1	$1.000\,00\,E-3$	$1.000\,00\,E-3$	$3.612\,73\,E-5$	$6.242\,80\,E-2$
$1.000\,00\,E+3$	1	1	$3.612\,73\,E-2$	$6.242\,80\,E+1$
$2.767\,99\,E+4$	$2.767\,99\,E+1$	$2.767\,99\,E+1$	1	$1.728\,00\,E+3$
$1.601\,85\,E+1$	$1.601\,85\,E-2$	$1.601\,85\,E-2$	$5.787\,04\,E-4$	1

5)   Surface tension (MT^{-2})

$N\,m^{-1} = J\,m^{-2}$	dyn cm^{-1} = erg cm^{-2}	kgf m^{-1}	lbf in^{-1}
1	1.000 00 E+3	1.019 72 E−1	5.710 15 E−3
1.000 00 E−3	1	1.019 72 E−4	5.710 15 E−6
9.806 65 E+0	9.806 65 E+3	1	5.599 74 E−2
1.751 27 E+2	1.751 27 E+5	1.785 80 E+1	1

6)   Force (MLT^{-2})

N	dyn	kgf	poundal	lbf
1	1.000 00 E+5	1.019 72 E−1	7.233 01 E+0	2.248 09 E−1
9.806 65 E+0	9.806 65 E+5	1	7.093 16 E+1	2.204 62 E+0
1.382 55 E−1	1.382 55 E+4	1.409 81 E−2	1	3.108 10 E−2
4.448 22 E+0	4.448 22 E+5	4.535 92 E−1	3.217 40 E+1	1

7)   Pressure ($ML^{-1}T^{-2}$)

Pa	bar	atm	kgf cm^{-2}	lbf in^{-2} (psi)
1	1.000 00 E−5	9.869 23 E−6	1.019 72 E−5	1.450 38 E−4
1.000 00 E+5	1	9.869 23 E−1	1.019 72 E+0	1.450 38 E+1
1.013 25 E+5	1.013 25 E+0	1	1.033 23 E+0	1.469 60 E+1
9.806 65 E+4	9.806 65 E−1	9.678 41 E−1	1	1.422 34 E+1
6.894 76 E+3	6.894 76 E−2	6.804 60 E−2	7.030 69 E−2	1

Pa	dyn cm^{-2}	mmHg (torr)	in Hg	lbf ft^{-2}
1	10	7.500 62 E−3	2.953 00 E−4	2.088 53 E−2
1.000 00 E−1	1	7.500 62 E−4	2.953 00 E−5	2.088 53 E−3
1.333 22 E+2	1.333 22 E+3	1	3.937 01 E−2	2.784 50 E+0
3.386 39 E+3	3.386 39 E+4	2.540 00 E+1	1	7.072 62 E+1
4.788 03 E+1	4.788 03 E+2	3.591 31 E−1	1.413 90 E−2	1

8)   Work, heat, energy (ML^2T^{-2})

J	erg	cal$_{th}$	Btu$_{th}$	kgf m
1	1.000 00 E+7	2.390 06 E−1	9.484 52 E−4	1.019 72 E−1
1.000 00 E−7	1	2.390 06 E−8	9.484 52 E−11	1.019 72 E−8
4.184 00 E+0	4.184 00 E+7	1	3.968 32 E−3	4.266 49 E−1
1.054 35 E+3	1.054 35 E+10	2.519 96 E+2	1	1.075 14 E+2
9.806 65 E+0	9.806 65 E+7	2.343 85 E+0	9.301 13 E−3	1

J	cal$_{IT}$	Btu$_{IT}$	kW h	HP h
1	2.388 46 E−1	9.478 13 E−4	2.777 78 E−7	3.725 06 E−7
4.186 80 E+0	1	3.968 30 E−3	1.163 00 E−6	1.559 61 E−6
1.055 06 E+3	2.519 97 E+2	1	2.930 72 E−4	3.930 16 E−4
3.600 00 E+6	8.598 45 E+5	3.412 13 E+3	1	1.341 02 E+0
2.684 52 E+6	6.411 87 E+5	2.544 42 E+3	7.457 00 E−1	1

9)   Specific enthalpy (L^2T^{-2})

J kg^{-1}	cal$_{th}$ g^{-1}	cal$_{IT}$ g^{-1}	Btu$_{th}$ lb^{-1}	Btu$_{IT}$ lb^{-1}
1	2.390 06 E−4	2.388 46 E−4	4.302 10 E−4	4.299 21 E−4
4.184 00 E+3	1	9.993 31 E−1	1.800 00 E+0	1.798 79 E+0
4.186 80 E+3	1.000 67 E+0	1	1.801 20 E+0	1.800 00 E+0
2.324 44 E+3	5.555 55 E−1	5.551 84 E−1	1	9.993 31 E−1
2.326 01 E+3	5.559 29 E−1	5.555 58 E−1	1.000 67 E+0	1

10)   Specific heat ($L^2T^{-2}\theta^{-1}$)

J kg^{-1} K^{-1}	cal$_{th}$ g^{-1} °C^{-1}	cal$_{IT}$ g^{-1} °C^{-1}	Btu$_{th}$ lb^{-1} °F^{-1}	Btu$_{IT}$ lb^{-1} °F^{-1}
1	2.390 06 E−4	2.388 46 E−4	2.390 06 E−4	2.388 46 E−4
4.184 00 E+3	1	9.993 31 E−1	1.000 00 E+0	9.993 31 E−1
4.186 80 E+3	1.000 67 E+0	1	1.000 67 E+0	1.000 00 E+0
4.184 00 E+3	1.000 00 E+0	9.993 31 E−1	1	9.993 31 E−1
4.186 80 E+3	1.000 67 E+0	1.000 00 E+0	1.000 67 E+0	1

11)   Power (ML^2T^{-3})

W	kgf m s^{-1}	lbf ft s^{-1}	HP	PS
1	1.019 72 E−1	7.375 62 E−1	1.341 02 E−3	1.359 62 E−3
9.806 65 E+0	1	7.233 02 E+0	1.315 09 E−2	1.333 33 E−2
1.355 82 E+0	1.382 55 E−1	1	1.818 18 E−3	1.843 40 E−3
7.457 00 E+2	7.604 02 E+1	5.500 00 E+2	1	1.013 87 E+0
7.354 99 E+2	7.500 00 E+1	5.424 76 E+2	9.863 20 E−1	1

12)   Viscosity ($ML^{-1}T^{-1}$)

Pa s	poise	kgf s m^{-2}	kgf h m^{-2}	lb h^{-1} ft^{-1}
1	1.000 00 E+1	1.019 72 E−1	2.832 55 E−5	2.419 09 E+3
1.000 00 E−1	1	1.019 72 E−2	2.832 55 E−6	2.419 09 E+2
9.806 65 E+0	9.806 65 E+1	1	2.777 78 E−4	2.372 32 E+4
3.530 39 E+4	3.530 39 E+5	3.600 00 E+3	1	8.540 38 E+7
4.133 79 E−4	4.133 79 E−3	4.215 28 E−5	1.170 91 E−8	1

Pa s	lbf s in^{-2}	lbf s ft^{-2}	lbf h in^{-2}	lbf h ft^{-2}
1	1.450 38 E−4	2.088 54 E−2	4.028 83 E−8	5.801 51 E−6
6.894 76 E+3	1	1.440 00 E+2	2.777 78 E−4	4.000 00 E−2
4.788 03 E+1	6.944 44 E−3	1	1.929 01 E−6	2.777 78 E−4
2.482 11 E+7	3.600 00 E+3	5.184 00 E+5	1	1.440 00 E+2
1.723 69 E+5	2.500 00 E+1	3.600 00 E+3	6.944 44 E−3	1

13) Thermal conductivity $(MLT^{-3}\theta^{-1})$

$W\,m^{-1}\,K^{-1}$	$cal_{th}\,s^{-1}$ $cm^{-1}\,°C^{-1}$	$kcal_{th}\,h^{-1}$ $m^{-1}\,°C^{-1}$	$Btu_{th}\,h^{-1}$ $ft^{-1}\,°F^{-1}$	$Btu_{th}\,in\,h^{-1}$ $ft^{-2}\,°F^{-1}$
1	2.390 06 E−3	8.604 21 E−1	5.781 76 E−1	6.938 11 E+0
4.184 00 E+2	1	3.600 00 E+2	2.419 09 E+2	2.902 91 E+3
1.162 22 E+0	2.777 78 E−3	1	6.719 68 E−1	8.063 62 E+0
1.729 58 E+0	4.133 79 E−3	1.488 17 E+0	1	1.200 00 E+1
1.441 31 E−1	3.444 82 E−4	1.240 14 E−1	8.333 33 E−2	1

14) Diffusivity $(L^2 T^{-1})$

$m^2\,s^{-1}$	stokes $(cm^2\,s^{-1})$	$m^2\,h^{-1}$	$in^2\,s^{-1}$	$ft^2\,h^{-1}$
1	1.000 00 E+4	3.600 00 E+3	1.550 00 E+3	3.875 01 E+4
1.000 00 E−4	1	3.600 00 E−1	1.550 00 E−1	3.875 01 E+0
2.777 78 E−4	2.777 78 E+0	1	4.305 56 E−1	1.076 39 E+1
6.451 60 E−4	6.451 60 E+0	2.322 58 E+0	1	2.500 00 E+1
2.580 64 E−5	2.580 64 E−1	9.290 31 E−2	4.000 00 E−2	1

15) Heat flux (MT^{-3})

$W\,m^{-2}$	$cal_{th}\,cm^{-2}\,s^{-1}$	$kcal_{th}\,m^{-2}\,h^{-1}$	$Btu_{th}\,ft^{-2}\,h^{-1}$
1	2.390 06 E−5	8.604 21 E−1	3.172 11 E−1
4.184 00 E+4	1	3.600 00 E+4	1.327 21 E+4
1.162 22 E+0	2.777 78 E−5	1	3.686 69 E−1
3.152 48 E+0	7.534 61 E−5	2.712 46 E+0	1

16) Heat transfer coefficient $(MT^{-3}\theta^{-1})$

$W\,m^{-2}\,K^{-1}$	$cal_{th}\,cm^{-2}$ $s^{-1}\,°C^{-1}$	$kcal_{th}\,m^{-2}$ $h^{-1}\,°C^{-1}$	$Btu_{th}\,ft^{-2}$ $h^{-1}\,°F^{-1}$
1	2.390 06 E−5	8.604 21 E−1	1.762 28 E−1
4.184 00 E+4	1	3.600 00 E+4	7.373 41 E+3
1.162 22 E+0	2.777 78 E−5	1	2.048 17 E−1
5.674 46 E+0	1.356 23 E−4	4.882 41 E+0	1

17) Temperature (θ)

$$T\,(K) = 273.15 + T\,(°C)$$
$$T\,(°C) = \frac{T\,(°F) - 32}{1.8}$$

A.3   Physical Properties of the Elements and Some Inorganic and Organic Compounds

TABLE A.3(a)
Elements and inorganic compounds

Name	Formula	Molecular weight	Specific gravity[a]	Melting point (°C)	Boiling point (°C)	Critical temperature (K)	Critical pressure (×10⁶ Pa)	Critical density (×10³ kg m⁻³)
Air	—	28.97	1.2928	—	−194	132.5	3.77	0.35
Argon	Ar	39.94	1.7828	−189.2	−185.7	151	4.86	0.531
Boron bromide	BBr$_3$	250.57	—	−46	96	573	—	0.90
Boron chloride	BCl$_3$	117.19	$(1.434)^0$	−107	12.5	452.0	3.87	—
Boron fluoride	BF$_3$	67.82	3.065	−127	−100.4	260.9	4.99	—
Bromine	Br$_2$	159.83	$(3.119)^{20}$	−7.2	58.78	584	10.3	0.848
Cyanogen	C$_2$N$_2$	52.02	2.3348	−34.4	−20.5	401	5.98	—
Carbon monoxide	CO	28.01	1.2501	−207	−192	134.2	3.55	0.311
Carbon dioxide	CO$_2$	44.01	1.9768	−56.6 (530 kPa)	−78.5 (sublimation)	304.3	7.40	0.460
Phosgene	COCl$_2$	98.92	$(1.434)^0$	−104	8.3	455.2	5.67	0.52
Carbon oxysulfide	COS	60.07	2.7149	−138.2	−50.2	378.2	6.18	—
Carbon disulfide	CS$_2$	76.13	$(1.2927)^0$	−108.6	46.3	546.2	7.70	0.441
Chlorine	Cl$_2$	70.91	3.2204	−101.6	−34.6	417.2	7.71	0.573
Deuterium	D$_2$	4.02	—	−254.4	−249.6	38.8	1.76	—
Heavy water	D$_2$O	20.03	$(1.10714)^{25}$	3.82	101.42	644.7	22.15	—
Fluorine	F$_2$	38.00	1.6354	−223	−187	118.2	2.53	—
Germanium tetrachloride	GeCl$_4$	214.43	$(1.8443)^{30}$	−49.5	84.0	550.2	3.85	—
Hydrogen	H$_2$	2.016	0.0898	−259.1	−252.7	33.3	1.30	0.0310
Hydrogen bromide	HBr	80.92	3.6445	−88.5	−67.0	363.2	8.51	—
Hydrogen cyanide	HCN	27.03	$(0.6876)^{20}$	−14	26	456.7	5.39	0.20
Hydrogen chloride	HCl	36.47	1.6394	−111	−85	324.6	8.27	0.42
Hydrogen fluoride	HF	20.01	$(0.987)^{15}$	−83	19.4	503.4	—	—
Hydrogen iodide	HI	127.93	5.7245	−50.8	−35.4	424.1	8.31	—
Water	H$_2$O	18.02	$(0.99708)^{25}$	0.0	100.0	647.4	22.13	0.323

TABLE A.3(a) – contd

Name	Formula	Molecular weight	Specific gravity[a]	Melting point (°C)	Boiling point (°C)	Critical temperature (K)	Critical pressure (×10^6 Pa)	Critical density (×10^3 kg m^-3)
Hydrogen peroxide	H_2O_2	34.02	$(1.438)^{20}$	−0.89	151.4	–	–	–
Hydrogen sulfide	H_2S	34.08	1.5392	−82.9	−59.6	373.6	9.01	2.86
Hydrogen selenide	H_2Se	81.22	–	−64	−42	411.2	8.92	–
Helium	He	4.00	0.1769	<−272.2	−268.9	5.3	0.229	0.0693
Mercury	Hg	200.61	$(13.546)^{20}$	−38.87	356.9	<1823	>20	4~5
Iodine	I_2	253.84	$(4.93)^{20}$	113.5	184.35	826.2	–	–
Krypton	Kr	83.70	3.6431	−169	−151.8	209.4	5.50	1.10
Nitrogen	N_2	28.02	1.2507	−209.86	−195.8	126.1	3.39	0.3110
Ammonia	NH_3	17.03	0.7708	−77.7	−33.4	405.6	11.30	0.235
Hydrazine	N_2H_4	32.05	$(1.011)^{15}$	1.4	113.5	653.2	14.69	–
Nitric oxide	NO	30.01	1.3401	−161	−151	179.2	6.59	0.52
Nitrous oxide	N_2O	44.02	1.9781	−102.3	−90.7	309.7	7.27	0.45
Nitrogen peroxide	N_2O_4	92.02	$(1.448)^{20}$	−9.3	21.3	431.2	10	1.785
Neon	Ne	20.18	0.8713	−248.67	−245.9	44.5	2.62	0.484
Oxygen	O_2	32.00	1.4289	−218.4	−183	154.4	5.04	0.430
Ozone	O_3	48.00	2.1415	−192.5	−111.9	261.1	5.53	0.326
Phosphine	PH_3	34.00	1.5293	−132.5	−85	324.2	6.48	0.30
Radon	Rn	222.0	9.73	−71	−62	377.2	6.28	–
Sulfur	S	32.06	–	120	444.6	1313	11.8	–
Sulfur dioxide	SO_2	64.06	2.9268	−77.5	−10.0	430.4	7.87	0.52
Sulfur trioxide	SO_3	80.06	$(1.97)^{20}$	16.83	44.6	491.5	8.47	0.630
Silicon chloride	$SiCl_4$	169.89	$(1.50)^{20}$	−70	57.6	506	–	–
Silicon fluoride	SiF_4	104.06	–	−95.7	–	271.7	5.07	–
Silane	SiH_4	32.09	1.44	−185	−112	269.7	4.86	–
Tin chloride	$SnCl_4$	260.53	(2.23)	−30.2	114.1	591.9	3.75	–
Xenon	Xe	131.30	5.7168	−140	−109.1	289.8	5.90	1.155

[a] The figures in parentheses are the specific gravities of the liquid at the temperature (°C) indicated in the superscript; others are densities (kg m^-3) of the gas at atmospheric pressure and 0°C.

TABLE A.3(b)
Organic compounds

Name	Formula	Molecular weight	Specific gravity[a]	Melting point (°C)	Boiling point (°C)	Critical temperature (K)	Critical pressure ($\times 10^6$ Pa)	Critical density ($\times 10^3$ kg m^{-3})
Methane	CH_4	16.04	0.7167	-182.5	-161.5	191.1	4.641	0.162
Ethane	C_2H_6	30.07	1.3567	-183.3	-88.6	305.6	4.894	0.203
Propane	C_3H_8	44.10	2.0200	-187.7	-42.1	370.0	4.257	0.220
n-butane	C_4H_{10}	58.12	2.5985	-138.4	-0.5	425.3	3.797	0.228
Isobutane	$(CH_3)_2CHCH_3$	58.12	$(0.5983)^{-13.6}$	-159.6	-11.7	408.2	3.648	0.221
n-pentane	C_5H_{12}	72.15	$(0.6262)^{20}$	-129.7	36.1	469.8	3.375	0.232
Isopentane	$(CH_3)_2CHC_2H_5$	72.15	$(0.6201)^{20}$	-160.0	28.0	461.0	3.33	0.234
Neopentane	$(CH_3)_4C$	72.15	$(0.613)^{0}$	-16.6	9.5	433.8	3.199	0.338
n-hexane	C_6H_{14}	86.18	$(0.6594)^{20}$	-95.3	68.7	507.9	3.034	0.234
n-heptane	C_7H_{16}	100.21	$(0.6838)^{20}$	-90.7	98.4	540.2	2.736	0.235
n-octane	C_8H_{18}	114.23	$(0.7025)^{20}$	-56.8	125.7	569.4	2.497	0.235
n-nonane	C_9H_{20}	128.25	$(0.718)^{20}$	-53.7	150.5	594.6	2.31	
n-decane	$C_{10}H_{22}$	142.28	$(0.730)^{20}$	-29.7	174.0	603.6	2.15	0.236
Cyclopentane	$(CH_2)_5$	70.13	$(0.745)^{20}$	-93.3	49	511.8	4.52	
Cyclohexane	$(CH_2)_6$	84.16	$(0.779)^{20}$	6.5	80	554.2	4.09	0.270
Ethylene	$CH_2{=}CH_2$	28.05	1.2644	-169	-103.9	282.9	5.12	0.22
Propylene	$CH_2{=}CHCH_3$	42.08	$(0.647)^{-79}$	-185	-47.0	365.5	4.56	0.233
Butadiene-(1,3)	$CH_2{=}CHCH{=}CH_2$	54.09	—	-108.9	-4.5	425	4.33	0.245
Acetylene	$CH{\equiv}CH$	26.04	1.1708	-81.5 (119 kPa)	-84	309.2	6.28	0.231

[a] The figures in parentheses are the specific gravities of the liquid at the temperature (°C) indicated in the superscript; others are densities (kg m^{-3}) of the gas at atmospheric pressure and 0°C.

A.4 Physical Properties of Some Gases

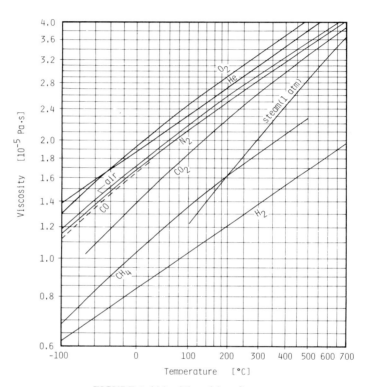

FIGURE A.4(a)   Viscosities of gases.

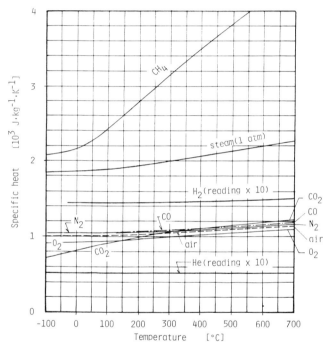

FIGURE A.4(b)   Specific heats of gases.

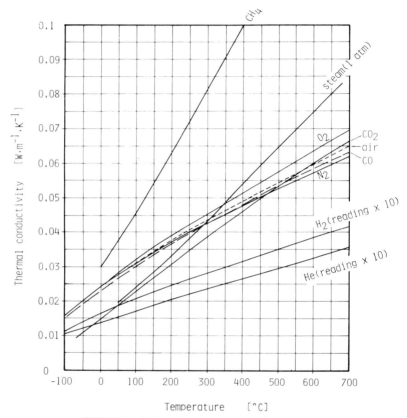

FIGURE A.4(c)   Thermal conductivities of gases.

A.5   Physical Properties of Some Liquids

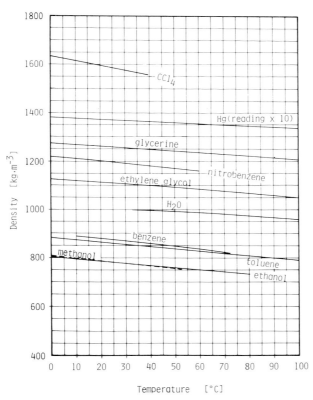

FIGURE A.5(a)   Densities of liquids.

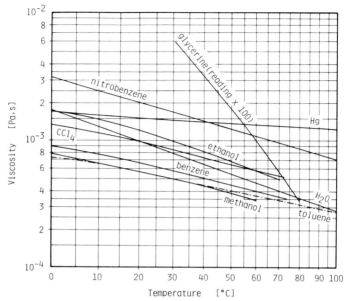

FIGURE A.5(b)   Viscosities of liquids.

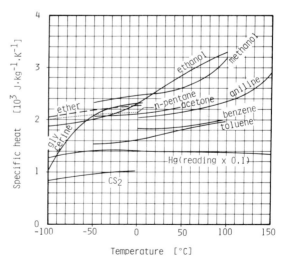

FIGURE A.5(c)   Specific heats of liquids.

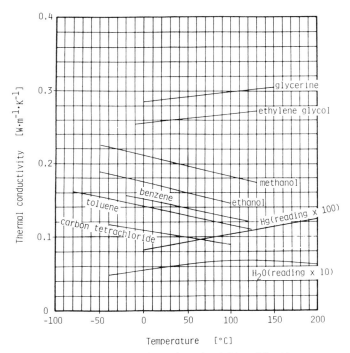

FIGURE A.5(d)   Thermal conductivities of liquids.

A.6 Physical Properties of Plastics

A.6.1 Polystyrene and Polyvinylchloride[a]

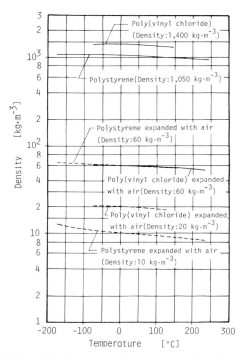

FIGURE A.6(a)   Densities of polystyrene and polyvinylchloride.

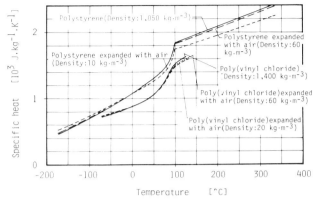

FIGURE A.6(b)   Specific heats of polystyrene and polyvinylchloride.

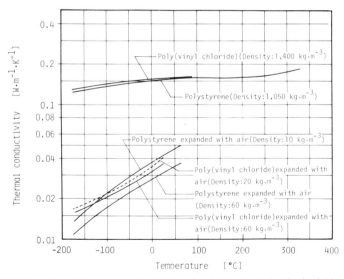

FIGURE A.6(c)   Thermal conductivities of polystyrene and polyvinylchloride.

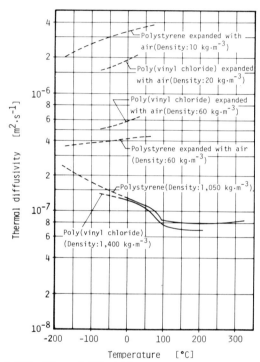

FIGURE A.6(d)   Thermal diffusivities of polystyrene and polyvinylchloride.

[a] *Source:* C. Y. Ho, P. D. Desai, K. Y. Wu, T. N. Havill and T. Y. Lee, Proceedings of the 7th Symposium on Thermophysical Properties, sponsored by ASME and NBS, Maryland, p. 198 (1977).

A.6.2   Polyethylene, Polypropylene and Polytetrafluoroethylene[a]

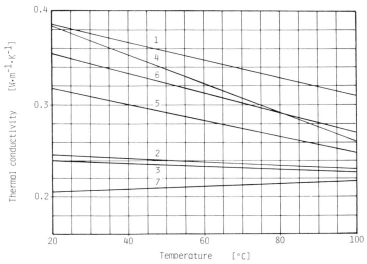

FIGURE A.6(e)   Thermal conductivities of polyethylene, polypropylene and poly-
tetrafluoroethylene.

[a] *Source:* K. W. Jackson and W. Z. Black, Proceedings of the 7th Symposium on
Thermophysical Properties, sponsored by ASME and NBS, Maryland, p. 141 (1977).

Sample	Material	Specifications–Use	Density $(kg\,m^{-3})$	Supplier
1	Ultrahigh molecular weight polyethylene	Used in high abrasion applications and cases where lubricity is important	949	Hercules Inc. Hifax® 1900 Polyethylene
2	Glass reinforced polypropylene with carbon black	40% continuous glass fiber, 59.5% polypropylene, 0.5% carbon black	1 088	GRTL Azdel® Laminate
3	Glass reinforced polypropylene without carbon black	40% continuous glass fiber, 60% polypropylene	1 170	GRTL Azdel® Laminate
4	Crosslinked Polyethylene EVA base, $CaCO_3$ filler	Ethylvinylacetate: 10% vinyl acetate filler; 28.5% by weight $CaCO_3$ filler	1 112	Union Carbide

5	Low density poly-ethylene	Unfilled crosslinked polyethylene	972	Union Carbide
6	Crosslinked poly-ethylene EVA base, SiO_2 filler	Ethylvinylacetate: 10% vinyl acetate filler; 28.5% by weight SiO_2 (silone treated) filler	1061	Union Carbide
7	Polytetrafluoro-ethylene	Standard Teflon® thermoplastic	2154	Dupont

A.7   Thermal Conductivities of Miscellaneous Solids

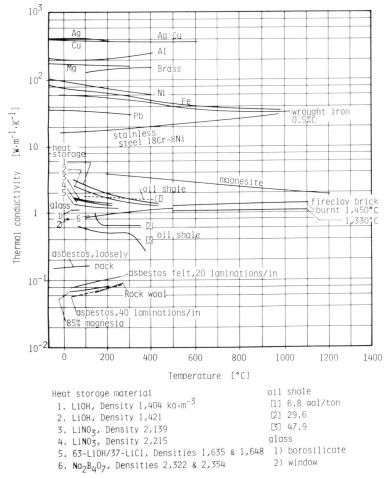

FIGURE A.7   Thermal conductivities of solids.

Permission has been granted to use material from the following sources: R. P. Tye, A. O. Desjarlais and J. G. Bourne, Proceedings of the 7th Symposium on Thermophysical Properties, sponsored by ASME and NBS, Maryland (1977) p. 189 for heat storage materials. R. Nottenburg, K. Rajeshwar, J. Dubow and R. Rosenvold, *ibid.* p. 396 for oil shales. Others are from A. I. Brown and S. M. Marco, *Introduction to Heat Transfer*, 3rd edn., McGraw-Hill, New York (1958); used with permission of the Publisher.

A.8   Prediction of Diffusion Coefficients in Binary Gas Systems

The following equations are recommended for predicting diffusion coefficients in binary gas systems:

D_{12} = diffusion coefficient (m² s⁻¹) between species 1 and 2

M_1, M_2 = molecular weights

P = total pressure (Pa)

T = temperature (K).

Hirschfelder-Bird-Spotz's formula: J. O. Hirschfelder, R. B. Bird and E. L. Spotz, *Trans. Amer. Soc. Mech. Eng.* **71**, 921 (1949).

$$D_{12} = \frac{BT^{1.5}}{P(\sigma_{12})^2 \Omega_D} \left(\frac{1}{M_1} + \frac{1}{M_2} \right)^{1/2}$$

where

$$B = 1.08 \times 10^{-4} \left[1 - 0.23 \left(\frac{1}{M_1} + \frac{1}{M_2} \right)^{1/2} \right]$$

$$\sigma_{12} = \frac{\sigma_1 + \sigma_2}{2} \text{ (nm)}$$

σ_1, σ_2 = collision diameters of species 1 and 2, respectively (nm).

Ω_D = collision integral for diffusion, function of kT/ϵ_{12}

ϵ_{12} = energy of molecular interaction: $\dfrac{\epsilon_{12}}{k} = \left[\left(\dfrac{\epsilon_1}{k} \right) \left(\dfrac{\epsilon_2}{k} \right) \right]^{1/2}$

k = Boltzmann constant.

For ϵ/k, σ, Ω_D, and k/ϵ_{12} see R. B. Bird, W. E. Stewart and E. N. Lightfoot, *Transport Phenomena*, John Wiley, New York (1960).

Andrussow's formula: L. Andrussow, *Z. Elektrochem.* **54**, 566 (1950); **55**, 51 (1951).

$$D_{12} = \frac{8.025 \times 10^{-7} T^{1.78} [1 + (M_1 + M_2)^{1/2}]}{P(V_1^{1/3} + V_2^{1/3})^2 (M_1 M_2)^{1/2}}$$

where V_1 and V_2 are the liquid molar volumes at normal boiling points (m³ mol⁻¹) of species 1 and 2, respectively.

Fujita's formula: S. Fujita, *Kagaku Kogaku* **28**, 251 (1964).

$$D_{12} = \frac{6.70 \times 10^{-8} T^{1.83}}{P \left[\left(\dfrac{T_c}{P_c} \right)_1^{1/3} + \left(\dfrac{T_c}{P_c} \right)_2^{1/3} \right]^3} \left(\frac{1}{M_1} + \frac{1}{M_2} \right)^{1/2}$$

where T_c and P_c are the critical temperature (K) and critical pressure (Pa), respectively; see Table A.3.

A.9   Data of Diffusion Coefficients in Binary Gas Systems

The figures times 10^{-4} give the diffusion coefficients ($m^2\ s^{-1}$) at the temperature (K) indicated in parentheses.

	air	Ar	CO	CO_2	D_2	H_2	H_2O	He
Air	air							
Ar		Ar						
CO		0.1880 (295.7) d	CO					
CO_2		0.1652 (317.2) d	0.1520 (296.1) d	CO_2				
D_2	0.5650 (296.8) d	0.5750 (296.8) d	0.5490 (295.7) d	0.4740 (295.7) d	D_2			
H_2	0.611 (273.2) a	0.8280 (287.9) d	0.7430 (295.6) d	0.6650 (298.2) d	1.2400 (288.2) d	H_2		
H_2O	0.2770 (312.6) d			0.2110 (328.6) d		1.0200 (307.3) d	H_2O	
He	0.6242 (276.2) d	0.8090 (323.2) d	0.7020 (295.6) d	0.6780 (323.2) d	1.2500 (295.1) d	1.1320 (298.2) d	1.4140 (498.2) d	He

	Kr	N_2	Ne	NH_3	N_2O	O_2
Kr	0.1190 (273.0) d					
N_2	0.1940 (293.0) d	0.2120 (295.8) d				
Ne	0.2710 (273.0) d	0.2400 (295.1) d	0.2230 (273.0) d			
NH_3	0.2320 (295.1) d	0.2480 (295.1) d	1.150 (303.9) b	0.2470 (295.1) d		
N_2O	0.1730 (300.0) d	0.185 (273.2) a	0.7050 (293.0) d	0.3030 (328.6) d	0.0531 (194.8) d	
O_2	0.178 (273.2) a	0.181 (273.2) a	0.697 (273.2) a	0.3180 (329.0) d	0.8090 (323.2) d	0.139 (273.2) a
SF_6	0.153 (273.2) a	0.0887 (296.8) d	0.6300 (296.8) d	0.3060 (295.7) d	0.3960 (286.2) d	0.1090 (296.6) d
Methane	0.2190 (289.0) d	0.220 (298.2) c	0.674 (273.0) a	0.3310 (328.8) d	1.0050 (373.0) d	0.216 (298.2) c
Ethane	0.1480 (298.0) d	0.535 (273.2) a	0.5420 (296.8) d	0.5370 (298.0) d	0.8310 (296.6) d	0.7260 (298.0) d

Compound										
Ethylene					0.1630 (291.2) d		0.2330 (328.5) d	0.486 (273.2) a		0.116 (273.2) a	
Propane		0.0860 (298.0) d								0.0863 (298.0) d	
n-butane					0.0960 (298.0) d			0.3610 (287.9) d			0.0663 (288.6) d
isobutane					0.0905 (298.0) d			0.277 (273.2) a			
n-hexane	0.0753 (288.6) d				0.0757 (288.6) d	0.5740 (417.0) d		0.2900 (288.7) d			0.0663 (288.6) d
2,3-dimethyl butane	0.0753 (288.4) d				0.0751 (288.7) d			0.3010 (288.8) d			0.0657 (288.9) d
Cyclohexane	0.0744 (288.6) d				0.0760 (288.6) d			0.3190 (288.5) d			0.0719 (288.9) d
Methyl cyclopentane	0.0742 (287.1) d				0.0760 (288.6) d			0.3180 (288.5) d			0.0731 (287.1) d
n-heptane					0.0740 (303.2) d	0.2650 (303.2) d		0.2830 (303.2) d	0.2180 (303.2) d		0.0658 (303.2) d

2,4-dimethyl pentane		0.0655 (303.2) d			0.2240 (303.2) d	0.2970 (303.3) d	0.2630 (303.2) d	0.0744 (303.1) d				
n-octane	0.0505 (273.2) a	0.0587 (303.2) d			0.2080 (303.2) d	0.2770 (303.2) d	0.2480 (303.2) d	0.0726 (303.1) d				0.0705 (303.1) d
2,2,4-trimethyl pentane		0.0599 (303.2) d			0.2120 (303.2) d	0.2920 (303.2) d	0.2530 (303.2) d	0.0713 (303.3) d				0.0705 (303.0) d
n-decane						0.3060 (364.1) d		0.0841 (363.6) d				
Benzene	0.0962 (298.2) d			0.0528 (273.2) a		0.4036 (311.3) d	0.3840 (298.2) d	0.1022 (311.3) d				0.1011 (311.3) d
Toluene	0.0920 (312.6) d	0.071 (273.2) a										
Methanol	0.132 (273.2) a			0.0879 (273.2) a		0.506 (273.2) a	1.0320 (423.2) d					
Ethanol	0.1350 (298.2) d			0.0685 (273.2) a		0.375 (273.2) a	0.4940 (298.2) d					
Propanol	0.0850 (273.2) a			0.0577 (273.2) a		0.315 (273.2) a	0.6760 (423.2) d					

2-propanol	0.0990 (299.1) d			0.6770 (423.2) d		
Butanol	0.0870 (299.1) d	0.0476 (273.2) a	0.2716 (273.2) a	0.5870 (423.2) d		
Ethylene oxide		0.0918 (298.0) d				0.0914 (298.0) d
Pyridine			0.4370 (317.9) d		0.1068 (317.9) d	0.1050 (318.3) d
Piperidine			0.4030 (314.7) d		0.0953 (314.9) d	0.0953 (315.0) d
Thiophene			0.4000 (302.2) d		0.0992 (302.1) d	0.0975 (302.1) d
Nitro-benzene	0.0855 (298.2) d			0.3720 (298.2) d		

The data are reproduced from (a) *International Critical Tables*, Vol. 5, McGraw-Hill, New York (1929), p. 62. (With permission of National Academy Press, Washington, D.C.). (b) R. Paul and I. B. Srivastava, *J. Chem. Phys.* **35**, 1621 (1961). (c) C. R. Mueller and R. W. Cahill, *J. Chem. Phys.* **40**, 651 (1964). (d) E. N. Fuller, P. D. Schettler and J. C. Giddings, *Ind. Eng. Chem.* **58** (No. 5), 19 (1966).

A.10   Diffusion Coefficients of Gases in Water

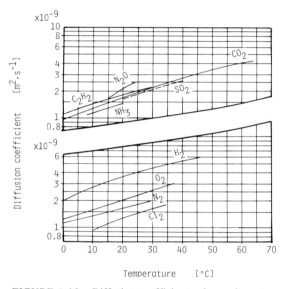

FIGURE A.10   Diffusion coefficients of gases in water.

Appendix B. Computer Programs (Fortran 77)

B.1 Prediction of Response Signal by the Method of Section 1.1.6.2; Calculation of Root-mean-square-errors for Construction of Two-dimensional Error Map

```
********************************************************************
********************************************************************
**                                                              **
**          ---------- MAIN PROGRAM ----------                  **
**                                                              **
**     Curve-fitting in time domain by Fourier analysis :       **
**     prediction of response signal from measured input signal;**
**     calculation of root-mean-square-errors between measured and **
**     predicted signals for the construction of two-dimensional **
**     error map.                                               **
**                                                              **
**     ----- Summary of Flow -----          MAIN :             **
**     Proc.1. normalization of measured input &    SUB. NORSIG  **
**           response signals                                   **
**           1.1. read title of experimental run    ========     **
**           1.2. read measured input signal      SUB. RDSIGL    **
**                read measured response signal    SUB. RDSIGL    **
**           1.3. normalize measured input signal  SUB. NORMLZ    **
**                normalize measured response signal SUB. NORMLZ   **
**           1.4. print input & response signals   SUB. PRINT    **
**     Proc.2. Fourier expansions of measured signals SUB. FOUEXP **
**           2.1. set expansion condition           ========     **
**           2.2. calc Fourier coef of input signal  SUB. FOUCOE  **
**                calc Fourier coef of response signl SUB. FOUCOE  **
**     Proc.3. reading of packed-bed parameter values SUB. RDPARM **
**     Proc.4. prediction of response curve        SUB. PRECUR   **
**           4.1. calc F. coef of predicted curve at  SUB. CLCCOE  **
**                various frequencies                           **
**                4.1.1. calc transfer function      SUB. TRANSF  **
**                4.1.2. calc Fourier coefficients    ========     **
**                       of predicted signal                    **
**           4.2. calc response curve & print       SUB. CLCCUR   **
**           4.3. calc root-mean-square-error & print ========    **
**     Proc.5. calculation of root-mean-square-errors SUB. RMSERR **
**           for construction of error map                      **
**           5.1. set two variable parameters       SUB. RDVARI  **
**                (norizontal & vertical parms)                 **
**           5.2. calc denominator of eqn(1.48)     ========     **
**           5.3. set horizontal parameter          ========     **
```

323

```
**        ---- Repeat calc by varying vertical   param  ----         **
**        ¦ -- Repeat calc by varying horizontal param ¦  --          **
**        ¦ ¦  5.4. calc F. coef of predicted curve at ¦ ¦SUB. CLCCOE **
**        ¦ ¦       various frequencies (same as 4.1)  ¦ ¦ SUB. TRANSF**
**        ¦ ¦  5.5. calc root-mean-square-error        ¦ ¦========    **
**        ¦ ----                                        ¦ --          **
**        ¦    5.6. print errors                        ¦ ========     **
**        ------                                          ----        **
**                                                                    **
**        ----- Variables in COMMON Blocks ----                       **
**        COMMON /SIGNAL/          data on input & response signals   **
**            TOIN,TORES : real;   starting times of signals          **
**            DTIN,DTRES : real;   time intervals                     **
**            NIN, NRES  : integer; number of data                    **
**            CIN, CRES  : real array; normalized signals             **
**        COMMON /FDATA /          data on Fourier expansion          **
**            TAU        : real;   half period                        **
**            NTERM      : integer; number of terms                   **
**            AOIN,AORES : real;   zeroth Fourier coefficients        **
**            AIN, ARES  : real array; Fourier cosine coefficients    **
**            BIN, BRES  : real array; Fourier sine coefficients      **
**            ( ---IN : input, ---RES : response )                    **
**        COMMON /PBPARM/          packed-bed parameters              **
**            PM         : real array; packed-bed parameter values    **
**                                 (see SUB. RDPARM)                  **
**                                                                    **
**********************************************************************
**********************************************************************
```

```
       COMMON       /PBPARM/ PM(20)
*                                      Proc.1. signal normalization
       CALL         NORSIG
*                                      Proc.2. Fourier expansion
       CALL         FOUEXP
*                                      Proc.3. parameter reading
       CALL         RDPARM(PM)
*                                      Proc.4. curve prediction
       CALL         PRECUR
*                                      Proc.5. r.m.s.e. calculation
       CALL         RMSERR

       STOP
       END
```

```
********************************************************************
********************************************************************
**                                                              **
**     SUBROUTINE    NORSIG                                      **
**                                                              **
********************************************************************
*                                                                *
*     Proc. 1.  normalization of measured input & response signals *
*               (title and measured signals are read from file '1') *
*                                                                *
*     source        : main program                              *
*     routines used : SUB. RDSIGL, NORMLZ & PRINT               *
*     data used     : none                                      *
*     returned data : COMMON /SIGNAL/                           *
*     working areas : TITLE,CDIN,CDRES,AIN,ARES                 *
*        TITLE      : character;   title of experimental run    *
*        CDIN,CDRES : real array;  measured input & response signals *
*        AIN, ARES  : real;        area of curves               *
*        ( ---IN : input, ---RES : response )                   *
*     format of input data from file '1':                       *
*        TITLE      : FORMAT( A80 )                              *
*                                                                *
********************************************************************


      SUBROUTINE    NORSIG
      COMMON        /SIGNAL/ TOIN, DTIN, NIN, CIN(200),
     $                       TORES,DTRES,NRES,CRES(200)
      DIMENSION     CDIN(200),CDRES(200)
      CHARACTER     TITLE*80
*                                          Proc.1.1. read title
      READ(1,100)   TITLE
      WRITE(6,600)  TITLE
*                                          Proc.1.2. read signals
      WRITE(6,610)
      CALL          RDSIGL(TOIN, DTIN, NIN, CDIN)
      CALL          RDSIGL(TORES,DTRES,NRES,CDRES)
*                                          Proc.1.3. normalize
      CALL          NORMLZ(DTIN, NIN, CDIN,  AIN, CIN)
      CALL          NORMLZ(DTRES,NRES,CDRES, ARES,CRES)
*                                          Proc.1.4. print
      CALL          PRINT(TOIN, DTIN, NIN, CDIN, AIN, CIN,
     $                    TORES,DTRES,NRES,CDRES,ARES,CRES)

      RETURN

******************* F O R M A T **************************************
  100 FORMAT( A80 )
  600 FORMAT( 1H1 / 1H0, 82(1H+) / 1H , 1H+, 80X, 1H+ / 1H , 1H+, A80,
     $ 1H+ / 1H , 1H+, 80X, 1H+ / 1H , 82(1H+) / )
  610 FORMAT( / 1H0, '***** MEASURED SIGNALS *****' )

      END
```

```
**************************************************************************
*                                                                        *
*       SUBROUTINE    RDSIGL(TO,DT,N,CD)                                  *
*                                                                        *
*       Proc.1.2. read measured signal from file '1'                     *
*                                                                        *
*       source       : SUB. NORSIG                                       *
*       routine used : none                                              *
*       data used    : none                                              *
*       returned data: TO,DT,N,CD                                        *
*               TO  : real;       starting time of signal               *
*               DT  : real;       time interval                         *
*               N   : integer;    number of data points                 *
*               CD  : real array; measured signal                       *
*       working area : CWORK                                             *
*               CWORK: real array; data in one block                     *
*       format of input data from file '1' :                            *
*               TO,DT: FORMAT( 2F10.0 )                                  *
*               CWORK: FORMAT( 10F8.0 )                                  *
*                                                                        *
*          Negative value of CWORK() indicates end of data.             *
*                                                                        *
**************************************************************************

        SUBROUTINE     RDSIGL(TO,DT,N,CD)
        DIMENSION      CD(200), CWORK(10)
*                                              +++++ read & check
        READ(1,100)    TO,DT
        WRITE(6,600) TO,DT
         IF( TO.LT.0.0 )  WRITE(6,610)
         IF( DT.LE.0.0 )  WRITE(6,620)
*                                              +++++ clear counter
        N = 0
*                                              +++++ read one block
     1 READ(1,110,END=90) (CWORK(I), I=1,10)
        WRITE(6,630)       (CWORK(I), I=1,10)
*                                              +++++ block division
        DO 10 I=1,10
         IF( CWORK(I).LT.0.0 )  RETURN
         N = N + 1
*                                              +++++ memory over ?
         IF( N.GT.200 )  GO TO 91
         CD(N) = CWORK(I)
    10 CONTINUE
        GO TO 1
*                                              print error messages
    90 WRITE(6,640)
        STOP
    91 WRITE(6,650)
        STOP
```

```
******************** F O R M A T *****************************************
  100 FORMAT( 2F10.0 )
  110 FORMAT( 10F8.0 )
  600 FORMAT( 1H0, 5X, 'T0 =', F10.5, 5X, 'DT =', F10.5 / )
  610 FORMAT( 1H0, 'STARTING TIME (T0) IS NEGATIVE.' / )
  620 FORMAT( 1H0, 'TIME INTERVAL (DT) IS NOT POSITIVE .' / )
  630 FORMAT( 1H , 10F8.2 )
  640 FORMAT( 1H0, 'END OF DATA CANNOT BE FOUND.  (SUB. RDSIGL)' / )
  650 FORMAT( 1H0, 'NUMBER OF DATA EXCEEDS 200. (SUB. RDSIGL)' / )

      END
```

```
****************************************************************
*                                                              *
*     SUBROUTINE     NORMLZ(DT,N,CD, AREA,CN)                  *
*                                                              *
*     Proc.1.3.  normalize measured signal                     *
*                                                              *
*     source       : SUB. NORSIG                               *
*     routine used : none                                      *
*     data used    : DT,N,CD     see SUB. RDSIGL              *
*     returned data: AREA,CN                                   *
*          AREA    : real;       area of curve                *
*          CN      : real array; normalized signal            *
*     method of integration : trapezoidal rule                 *
*                                                              *
****************************************************************

      SUBROUTINE     NORMLZ(DT,N,CD, AREA,CN)
      DIMENSION      CD(200), CN(200)
*                                              +++++ calc AREA
      AREA = 0.0
      DO 10 I=1,N
      AREA = AREA + CD(I)
   10 CONTINUE
      AREA = AREA*DT
*                                              +++++ check AREA
      IF( AREA.EQ.0.0 )  THEN
         WRITE(6,600)
         DO 20 I=1,N
           CN(I) = CD(I)
   20    CONTINUE
         RETURN
       ELSE IF( AREA.LT.0.0 )  THEN
         WRITE(6,610)
      END IF
*                                              +++++ normalize
      DO 30 I=1,N
      CN(I) = CD(I)/AREA
   30 CONTINUE

      RETURN

******************* F O R M A T ********************************
  600 FORMAT( 1H0, 'AREA OF CURVE IS ZERO (RETURNED SIGNAL IS NOT NORMAL
     $IZED).' / )
  610 FORMAT( 1H0, 'AREA OF CURVE IS NEGATIVE.' / )

      END
```

```
*****************************************************************************
*                                                                          *
*      SUBROUTINE    PRINT(TOIN, DTIN, NIN, CDIN, AIN, CIN,                 *
*      $                    TORES,DTRES,NRES,CDRES,ARES,CRES)               *
*                                                                          *
*      Proc.1.4.  print input & response signals (measured & normalized)*  *
*                                                                          *
*      source       : SUB. NORSIG                                          *
*      routine used : none                                                 *
*      data used    : TOIN, DTIN, NIN, CDIN, AIN, CIN,                     *
*                     TORES,DTRES,NRES,CDRES,ARES,CRES  see SUB. NORSIG     *
*      returned data: none                                                 *
*                                                                          *
*****************************************************************************

       SUBROUTINE    PRINT(TOIN, DTIN, NIN, CDIN, AIN, CIN,
       $                    TORES,DTRES,NRES,CDRES,ARES,CRES)
       DIMENSION     CDIN(200),CIN(200),CDRES(200),CRES(200)
*                                                  +++++ heading
       WRITE(6,600)
       WRITE(6,610)  TOIN, DTIN, NIN, AIN
       WRITE(6,620)  TORES,DTRES,NRES,ARES
*                                                  print input & response
  .    WRITE(6,630)
*                                                  +++ set number of data
       M = MAX(NIN,NRES)
       DO 10 I=1,M
*                                                  +++++ set times
       TIN = TOIN + FLOAT(I-1)*DTIN
       TRES= TORES+ FLOAT(I-1)*DTRES
*                                                  +++++ print signals
       IF( I.GT.NIN )  THEN
           WRITE(6,640) I, TRES,CDRES(I),CRES(I)
         ELSE IF( I.GT.NRES )  THEN
           WRITE(6,650) I, TIN,CDIN(I),CIN(I)
         ELSE
           WRITE(6,660) I, TIN,CDIN(I),CIN(I), TRES,CDRES(I),CRES(I)
       END IF
  10 CONTINUE

       RETURN

******************* F O R M A T  ***************************************
  600 FORMAT( // 1H0,'***** INPUT AND RESPONSE SIGNALS *****' / 1H0, 10X
     $,6X,'STARTING TIME',5X,'TIME INTERVAL',5X,'NO OF DATA',5X,'AREA' )
  610 FORMAT( 1H0, 5X,' INPUT  ',4X,F10.4,7X,F10.4,8X,I5,5X,E13.5 )
  620 FORMAT( 1H0, 5X,'RESPONSE',4X,F10.4,7X,F10.4,8X,I5,5X,E13.5 )
  630 FORMAT( 1H0, 15X, '======== INPUT ========', 15X,'======= RESPONSE
     $ =======' / 1H ,4X,'N',2(7X,'TIME       READING  (NORMALIZED)'))
  640 FORMAT( 1H , I5, 38X, 3X, F10.4, F10.1, 3X, F10.5 )
  650 FORMAT( 1H , I5, 3X, F10.4, F10.1, 3X, F10.5 )
  660 FORMAT( 1H , I5, 2(3X, F10.4, F10.1, 3X, F10.5, 2X) )

       END
```

```
*********************************************************************
*********************************************************************
**                                                                 **
**     SUBROUTINE    FOUEXP                                         **
**                                                                 **
*********************************************************************
*                                                                   *
*     Proc. 2.  Fourier expansions of measured input & response signals*
*               (half period & number of terms are read from file '2') *
*                                                                   *
*     source       : main program                                   *
*     routine used : SUB. FOUCOE                                     *
*     data used    : COMMON /SIGNAL/         see main program       *
*     returned data: COMMON /FDATA /         see main program       *
*     format of input data from file '2' :                          *
*        TAU,NTERM: FORMAT( F10.0,I10 )                             *
*        TAU      : real;         half period                       *
*        NTERM    : integer;      number of terms                   *
*                                                                   *
*********************************************************************

      SUBROUTINE    FOUEXP
      COMMON        /SIGNAL/ TOIN, DTIN, NIN, CIN(200),
     $                       TORES,DTRES,NRES,CRES(200)
      COMMON        /FDATA / TAU, NTERM, AOIN, AIN(200), BIN(200),
     $                       AORES,ARES(200),BRES(200)

*                                             Proc.2.1. set paramters
      READ(2,200)   TAU,NTERM
      WRITE(6,600)  TAU,NTERM
*                                         +++++ check parameters
      IF( TAU.LT.(TORES+DTRES*FLOAT(NRES))/2.0 ) THEN
         TAU = (TORES + DTRES*FLOAT(NRES))/2.0 * 1.5
         WRITE(6,610)  TAU
      END IF
      IF( NTERM.LT.1 .OR. NTERM.GT.IFIX(TAU/DTRES+0.5) )  THEN
         NTERM = IFIX(TAU/DTRES + 0.5)
         WRITE(6,620)  NTERM
      END IF
      IF( NTERM.GT.200 )  THEN
        NTERM = 200
        WRITE(6,620)  NTERM
      END IF
*                                             Proc.2.2. calc F. coef
      WRITE(6,630)
      CALL          FOUCOE(TOIN, DTIN, NIN, CIN, TAU,NTERM,AOIN, AIN, BIN )
      WRITE(6,640)
      CALL          FOUCOE(TORES,DTRES,NRES,CRES,TAU,NTERM,AORES,ARES,BRES)

      RETURN
```

```
******************** F O R M A T *****************************************
 200 FORMAT( F10.0, I10 )
 600 FORMAT( // 1H0, '***** FOURIER EXPANSION *****' /  1H0, 5X, 'HALF
    $PERIOD =', E13.5, 5X, 'NO OF TERMS =', I5 )
 610 FORMAT( 1H0, 'HALF PERIOD (TAU) IS REPLACED BY', E15.7 )
 620 FORMAT( 1H0, 'NUMBER OF TERMS (NTERM) IS REPLACED BY', I5 )
 630 FORMAT( 1H0, 5X, '+++++ CONVERSION CHECK (INPUT) +++++' )
 640 FORMAT( 1H0, 5X, '+++++ CONVERSION CHECK (RESPONSE) +++++' )

    END
```

```
*************************************************************************
*                                                                       *
*       SUBROUTINE    FOUCOE(TO,DT,NT,CN,TAU,NTERM, AO,A,B)             *
*                                                                       *
*       Proc.2.2.   calculate Fourier coefficients of signals (eqns(1.41a)*
*                   & (1.41b) for input; eqns(1.46a) & (1.46b) for      *
*                   response signal)                                    *
*                                                                       *
*       source        : SUB. FOUEXP                                     *
*       data used     : TO,DT,NT,CN,TAU,NTERM;  see SUB. RDSIGL & FOUEXP *
*             CN(N) : real;   signal value at n-th point                *
*       returned data: AO,A,B                                           *
*             AO    : real;   zeroth Fourier coefficient                *
*             A(N)  : real;   n-th Fourier cosine coefficient           *
*             B(N)  : real;   n-th Fourier sine coefficient             *
*       working areas: PI,COTEST,SQ,SQTEST,RAT                          *
*             PI    : real;   circle circumference-to-diameter ratio    *
*             COTEST: real;   calculated signal at t=0 by Fourier series*
*             SQ    : real;   integral of signal squared                *
*             SQTEST: real;   integral of signal squared, predicted in  *
*                             terms of Fourier coefficients             *
*             RAT   : real;   ratio of SQTEST to SQ                     *
*             (COTEST & SQTEST are used to check convergence of         *
*              Fourier series.)                                         *
*       integration formula: trapezoidal rule                          *
*                                                                       *
*************************************************************************

        SUBROUTINE    FOUCOE(TO,DT,NT,CN,TAU,NTERM, AO,A,B)
        DIMENSION     CN(200), A(200),B(200)
        DATA          PI / 3.141593 /
*                                               +++++ calc AO & SQ
        AO = 0.0
        SQ = 0.0
        DO 10 I=1,NT
        AO = AO + CN(I)
        SQ = SQ + CN(I)**2
     10 CONTINUE
        AO = AO*DT/TAU
        SQ = SQ*DT
*                                               +++++ heading
        WRITE(6,600)
*                                               +++++ initial set
        COTEST = AO/2.0
        SQTEST = 2.0*(AO/2.0)**2 * TAU
*                                               +++++ calc A(N) & B(N)
        DO 20 N=1,NTERM
        X = 0.0
        Y = 0.0
*                                               +++++ save SV
        SV = FLOAT(N)*PI/TAU
        DO 30 I=1,NT
          Z = SV*(TO + FLOAT(I-1)*DT)
          X = X + CN(I)*COS(Z)
          Y = Y + CN(I)*SIN(Z)
```

```
   30    CONTINUE
         A(N) = X*DT/TAU
         B(N) = Y*DT/TAU
*                                               +++ clc COTEST & SQTEST
         COTEST = COTEST + A(N)
         SQTEST = SQTEST + (A(N)**2 + B(N)**2)*TAU
*                                               +++++ print check data
         IF( MOD(N,10).EQ.0 )  THEN
           RAT = SQTEST/SQ
           WRITE(6,610)   N,RAT,COTEST
         END IF
   20  CONTINUE

       RETURN

******************** F O R M A T ************************************
   600 FORMAT( 1H0, 10X, 'NO OF TERMS', 8X, 'RATIO', 7X, 'VALUE AT T=0' )
   610 FORMAT( 1H , 10X, I5, 10X, F10.5, 5X, F10.5 )

       END
```

```
**********************************************************************
*                                                                    *
*      SUBROUTINE   RDPARM(PM)                                        *
*                                                                    *
*      Proc. 3.  reading of packed-bed parameter values from file '3' *
*                                                                    *
*      source       : main program                                   *
*      routine used : none                                           *
*      data used    : none                                           *
*      returned data: PM                                             *
*          PM     : real array; packed-bed parameters                *
*          PM(1) : 'L  ' ;    length of bed                          *
*          PM(2) : 'U  ' ;    interstitial flow rate                 *
*          PM(3) : 'SA ' ;    surface area per unit volume           *
*          PM(4) : 'EB ' ;    bed void fraction                      *
*          PM(5) : 'R  ' ;    particle radius                        *
*          PM(6) : 'EP ' ;    intraparticle void fraction            *
*          PM(7) : 'RHOP';    particle density                       *
*          PM(8) : 'DV ' ;    molecular diffusivity                  *
*          PM(9) : 'DAX ';    axial dispersion coefficient           *
*          PM(10): 'DE ' ;    intraparticle effective diffusivity    *
*          PM(11): 'SH ' ;    Sherwood number (= 2*SKF*R/DV)          *
*                  SKF  ;    particle-to-fluid transfer coefficient  *
*          PM(12): 'SKA ';    adsorption rate constant;              *
*                            set SKA=0 when SKA is infinitely large  *
*          PM(13): 'KA  ';    adsorption equilibrium constant        *
*      working areas: IP,PNAME                                       *
*          IP     : integer;  number of parameters                   *
*                            (IP=13 in this program)                 *
*          PNAME(I): character; parameter symbol for PM(I)           *
*      format of input data from file '3' :                          *
*          PM(1) to PM(13): FORMAT( 8F10.0 )                         *
*                                                                    *
**********************************************************************


      SUBROUTINE   RDPARM(PM)
      DIMENSION    PM(20)
      CHARACTER*4  PNAME(20)
      DATA         PNAME / 'L  ','U  ','SA ','EB ','R  ','EP ',
     $                     'RHOP','DV ','DAX ','DE ','SH ','SKA ','KA  ' /

      IP = 13
*                                           +++++ read param values
      READ(3,300)  (PM(I), I=1,IP)
*                                           +++++ print
      WRITE(6,600)
      WRITE(6,610) (I,PNAME(I),PM(I), I=1,IP)

      RETURN

****************** F O R M A T ************************************
  300 FORMAT( 8F10.0 )
  600 FORMAT( // 1H0, '***** PACKED-BED PARAMETERS *****' / )
  610 FORMAT(( 1H, 4(2X, I2, ') ', A4, '=', E9.3) ))

      END
```

```
************************************************************************
************************************************************************
**                                                                  **
**      SUBROUTINE    PRECUR                                         **
**                                                                  **
************************************************************************
*                                                                    *
*      Proc. 4.  prediction of response curve & root-mean-square-error *
*                                                                    *
*      source       : main program                                   *
*      routines used: SUB. CLCCOE & CLCCUR                            *
*      data used    : COMMON /SIGNAL/, /FDATA / & /PBPARM/            *
*      returned data: none                                           *
*      working areas: AOCLC,ACLC,BCLC,NT,ERR                         *
*              AOCLC,ACLC,BCLC,NT see SUB. CLCCOE                     *
*              ERR  : real;     root-mean-square-error between        *
*                               measured & predicted curves, eqn(1.48)*
*      method of calculation : eqns(1.42), (1.45) & (1.48)            *
*                                                                    *
************************************************************************

       SUBROUTINE    PRECUR
       COMMON        /SIGNAL/ TOIN, DTIN, NIN, CIN(200),
      $                       TORES,DTRES,NRES,CRES(200)
       COMMON        /FDATA / TAU, NTERM, AOIN, AIN(200), BIN(200),
      $                       AORES,ARES(200),BRES(200)
       COMMON        /PBPARM/ PM(20)
       DIMENSION     ACLC(200),BCLC(200)
*                                               +++++ heading
       WRITE(6,600)
*                                               Proc.4.1. calc F. coeff
       CALL          CLCCOE(PM,TAU,NTERM,NT,AOIN,AIN,BIN,AOCLC,ACLC,BCLC)
*                                               Proc.4.2. calc response
       CALL          CLCCUR(TAU,NT,AOCLC,ACLC,BCLC,0.0,2.0*TAU,DTRES)
*                                               Proc.4.3. calc r.m.s.e.
       X = 2.0*(AORES/2.0 - AOCLC/2.0)**2
       Y = 2.0*(AORES/2.0)**2
       DO 40 I=1,NT
       X = X + (ARES(I)-ACLC(I))**2 + (BRES(I)-BCLC(I))**2
       Y = Y + ARES(I)**2 + BRES(I)**2
   40 CONTINUE
       ERR = SQRT(X/Y)
       WRITE(6,610)  ERR

       RETURN

******************** F O R M A T ************************************
  600 FORMAT( // 1H0, '***** CALCULATION OF RESPONSE CURVE *****' )
  610 FORMAT( 1H0, 5X, 'ROOT-MEAN-SQUARE-ERROR =', F8.4 )

       END
```

```
*****************************************************************
*                                                               *
*     SUBROUTINE    CLCCOE(PM,TAU,NTERM,NT,AOIN, AIN, BIN,       *
*     $                                 AOCLC,ACLC,BCLC)         *
*                                                               *
*     Proc.4.1.  calculate Fourier coefficients of predicted curve*
*                at various frequencies                         *
*                                                               *
*     source       : SUB. PRECUR (Proc.4), or SUB. RMSERR (Proc.5)*
*     routine used : SUB. TRANSF                                *
*     data used    : PM,TAU,NTERM,AOIN,AIN,BIN   see SUB. PRECUR *
*     returned data: AOCLC,ACLC,BCLC                            *
*          AOCLC : real;     zeroth coefficient                 *
*          ACLC(N): real;    n-th cosine coefficient            *
*          BCLC(N): real;    n-th sine coefficient              *
*          NT    : integer;  minimum number of terms needed for *
*                            convergence of Fourier series for  *
*                            predicted signal                   *
*     working areas: PI,W                                       *
*          PI    : real;     circle circumference-to-diameter ratio*
*          W     : real;     frequency                          *
*     method of calculation: eqn(1.45)                          *
*                                                               *
*****************************************************************

      SUBROUTINE    CLCCOE(PM,TAU,NTERM,NT,AOIN, AIN, BIN,
      $                                 AOCLC,ACLC,BCLC)
      DIMENSION     PM(20), AIN(200),BIN(200), ACLC(200),BCLC(200)
      DATA          PI / 3.141593 /
*                                          +++++ zero frequency
      CALL          TRANSF(0, 0.0,PM, AN,BN)
      AOCLC = AOIN*AN
*                                          Repeat clc varying freq
      DO 10 N=1,NTERM
       W = FLOAT(N)*PI/TAU
*                                          Proc.4.1.1. calc transf
      CALL          TRANSF(1, W,PM, AN,BN)
*                                          Proc.4.1.2. calc coeffi
       ACLC(N) = AIN(N)*AN + BIN(N)*BN
       BCLC(N) = BIN(N)*AN - AIN(N)*BN
*                                          +++++ check conversion
      IF( AN**2+BN**2.LT.1.0E-8 )  THEN
         NT = N
         RETURN
      END IF
  10 CONTINUE
      NT = NTERM

      RETURN
      END
```

```
*********************************************************************
*                                                                   *
*     SUBROUTINE     TRANSF(ICTL, W,PM, AN,BN)                       *
*                                                                   *
*     Proc.4.1.1.  calculate transfer function for adsorption system *
*                                                                   *
*     source          : SUB. CLCCOE                                 *
*     data used       : ICTL,W,PM                                   *
*             ICTL : integer;   =0, initial call for parameter set  *
*                                nonzero, otherwise                 *
*             PM   : real array; packed-bed parameters, see SUB.RDPARM*
*     returned data : AN,BN                                         *
*             AN   : real;  real part of transfer function          *
*             BN   : real;  imaginary part of transfer function     *
*     save variables: D1,D2,D3,D4,D5,D6,D7,D8,D9                    *
*             D1 = L*U/(2.0*DAX);D2 = 4.0*DAX/U**2;                 *
*             D3 = SA*DE/(EE*R); D4 = DE/(R*SKF);  D5 = 1/DE;       *
*             D6 = RHOP*KA;     D7 = KA/SKA;    D8 = R;  D9 = EP    *
*     working areas: CS,CPH1,COTH,CQ,CF                             *
*             CS   : complex*16;  Laplace operator                  *
*             CPH1: complex*16;  phi-a in eqn(1.61c)               *
*             CEX : complex*16;  exp(-2*phi-a)                      *
*             COTH: complex*16;  cotn(phi-a)                        *
*             CQ  : complex*16;  q in eqn(1.61b)                    *
*             CF  : complex*16;  transfer function, eqn(1.61)       *
*     method of calculation: eqn(1.61)                             *
*                                                                   *
*        For other forms of transfer function, only IP and PNAME    *
*        (in Proc.3) and Proc.4.1.1 need to be alterd.             *
*                                                                   *
*********************************************************************

      SUBROUTINE     TRANSF(ICTL, W,PM, AN,BN)
      DIMENSION      PM(20)
      REAL*8         D1,D2,D3,D4,D5,D6,D7,D8,D9
      SAVE           D1,D2,D3,D4,D5,D6,D7,D8,D9
      COMPLEX*16     CS,CPHI,CEX,COTH,CQ,CF
*                                                 +++++ initial call ?
      IF( ICTL.EQ.0 )  THEN
*                                                 +++++ set save variables
        D1 = PM(1)*PM(2)/(2.0*PM(9))
        D2 = 4.0*PM(9)/PM(2)**2
        D3 = PM(3)*PM(10)/(PM(4)*PM(5))
          SKF = PM(11)*PM(8)/(2.0*PM(5))
        D4 = PM(10)/(PM(5)*SKF)
        D5 = 1.0/PM(10)
        D6 = PM(7)*PM(13)
        IF( PM(12).EQ.0.0 )  THEN
            D7 = 0.0
          ELSE
            D7 = PM(13)/PM(12)
        END IF
        D8 = PM(5)
        D9 = PM(6)
      END IF
*                                                 +++++ zero frequency ?
```

```
      IF( W.EQ.0.0 ) THEN
        AN = 1.0
        BN = 0.0
        RETURN
      END IF
*                                          +++++ calc transfer func
        CS  = CMPLX(0.0,W)
        CPHI= D8*CDSQRT(CS*D5*(D9 + D6/(D7*CS+1.0)))
        CEX = CDEXP(-2.0*CPHI)
        COTH= (1.0+CEX)/(1.0-CEX)
        CQ  = D3/(D4 + 1.0/(CPHI*COTH-1.0))
        CF  = CDEXP(D1*(1.0 - CDSQRT(1.0+D2*(CS+CQ))))
      AN = CF
      BN = DIMAG(CF)

      RETURN
      END
```

```
*****************************************************************************
*                                                                           *
*      SUBROUTINE   CLCCUR(TAU,NT,A0,A,B,T1,T2,DT)                           *
*                                                                           *
*      Proc.4.2.  calculate response curve (in time range from T1           *
*                 to T2) with Fourier coefficients and print                *
*                                                                           *
*      source       : SUB. PRECUR                                           *
*      routine used : none                                                  *
*      data used    : TAU,NT,A0,A,B,T1,T2,DT                                *
*           TAU,NT: see main program & SUB. PRECUR                          *
*           A0    : real;   zeroth Fourier coefficient                      *
*           A(N)  : real;   n-th Fourier cosine coefficient                 *
*           B(N)  : real;   n-th Fourier sine coefficient                   *
*           T1,T2 : real;   time range                                      *
*           DT    : real;   time interval                                   *
*      returned data: none                                                  *
*      working areas: PI,DTW,T,C,IN,ERR                                     *
*           PI    : real;   circle circumference-to-diameter ratio          *
*           DTW   : real;   time interval, working area of DT               *
*           T(N)  : real;   time at n-th point                              *
*           C(N)  : real;   signal value at n-th point                      *
*           IN    : integer; number of data points                         *
*      method of calculation: eqn(1.42)                                     *
*                                                                           *
*****************************************************************************

      SUBROUTINE   CLCCUR(TAU,NT,A0,A,B,T1,T2,DT)
      DIMENSION    A(200),B(200),C(500),T(500)
      DATA         PI / 3.141593 /
*                                             +++ set number of points
      WRITE(6,600)  T1,T2,DT
      IF( DT.EQ.0 )  THEN
         IN = 1
         DTW= DT
       ELSE
         IN = IFIX((T2-T1)/DT + 0.5)
         IF( IN.GE.0 )  THEN
            IN = IN + 1
            DTW= DT
          ELSE
            IN = -IN + 1
            DTW= -DT
         END IF
         IF( IN.GT.500 )  THEN
           IN = 500
           WRITE(6,610)  IN
         END IF
      END IF
*                                             +++++ initial set
      DO 10 I=1,IN
        T(I) = T1 + FLOAT(I-1)*DTW
        C(I) = A0/2.0
   10 CONTINUE
*                                             +++ calc response curve
```

```
      DO 20 N=1,NT
*                                                +++++ save SV
        SV = FLOAT(N)*PI/TAU
        DO 30 I=1,IN
          Z = SV*T(I)
          C(I) = C(I) + (A(N)*COS(Z) + B(N)*SIN(Z))
   30   CONTINUE
   20 CONTINUE
*                                                +++++ print
      WRITE(6,620)
      WRITE(6,630)  (T(I),C(I), I=1,IN)

      RETURN

******************** F O R M A T ****************************************
  600 FORMAT( 1H0, 5X, 'TIME RANGE :', F7.2, ' TO ', F7.2, 5X, 'INTERVAL
     $ =', F7.3 )
  610 FORMAT( 1H0, 'NUMBER OF DATA (IN) IS REPLACED BY', I5 / )
  620 FORMAT( 1H0, 2X, 5(' TIME  RESPONSE ') )
  630 FORMAT(( 1H , 5(F7.2, F9.5) ))

      END
```

```
********************************************************************
********************************************************************
**                                                               **
**     SUBROUTINE    RMSERR                                       **
**                                                               **
********************************************************************
*                                                                 *
*     Proc. 5. calculation of root-mean-square-errors between measured *
*              and predicted signals for construction of error map *
*                                                                 *
*     source       : main program                                 *
*     routine used : SUB. RDVARI & CLCCOE                         *
*     data used    : COMMON /FDATA / & /PBPARM/    see main program *
*     returned data: none                                         *
*     working areas: NP1,N1,VS1,DP1,NP2,N2,VS2,DP2   see SUB. RDVARI *
*                    PMW,PM1,ERR,ASQ,IQ                           *
*         PMW  : real array;  parameter set with varying horizontal *
*                             and vertical parameters             *
*         PM1  : real array;  horizontal parameter values         *
*         ERR  : real array;  root-mean-square-errors for PM1 values *
*         ASQ  : real;        denominator of eqn(1.48)            *
*         IQ   : character;   character for heading                *
*     method of calculation : eqn(1.48)                           *
*                                                                 *
********************************************************************


      SUBROUTINE    RMSERR
      COMMON        /FDATA / TAU, NTERM, AOIN, AIN(200), BIN(200),
     $                       AORES,ARES(200),BRES(200)
      COMMON        /PBPARM/ PM(20)
      DIMENSION     PMW(20),PM1(11),ERR(11)
      DIMENSION     ACLC(200),BCLC(200)
      CHARACTER     IQ*10
      DATA          IQ / ' =========' /

      WRITE(6,600)
*                                           Proc.5.1. set two param
      CALL          RDVARI(NP1,N1,VS1,DP1, NP2,N2,VS2,DP2)

*                                           Proc.5.2. calc ASQ
      ASQ = 2.0*(AORES/2.0)**2
      DO 10 N=1,NTERM
      ASQ = ASQ + ARES(N)**2 + BRES(N)**2
   10 CONTINUE
      ASQ = SQRT(ASQ)
*                                           ++++ initial set of PMW
      DO 20 I=1,20
      PMW(I) = PM(I)
   20 CONTINUE
*                                           Proc.5.3. set hor param
      DO 30 I1=1,N1+1
      PM1(I1) = VS1 + DP1*FLOAT(I1-1)
   30 CONTINUE
*                                           +++++  heading
```

```
      WRITE(6,610)  (PM1(I1), I1=1,N1+1)
      WRITE(6,620)  (IQ, I1=1,N1+1)
*                                            Repeat calc verticl prm
      DO 40 I2=1,N2+1
*                                            +++++ set ver prm value
        PMW(NP2) = VS2 + DP2*FLOAT(I2-1)
*                                            Repeat calc horizon prm
        DO 50 I1=1,N1+1
*                                            +++++ set hor prm value
          PMW(NP1) = PM1(I1)
*                                            Proc.5.4. calc F. coeff
          CALL      CLCCOE(PMW,TAU,NTERM,NT,AOIN,AIN,BIN,AOCLC,ACLC,BCLC)

*                                            Proc.5.5. calc r.m.s.e.
          X = 2.0*(AORES/2.0 - AOCLC/2.0)**2
          DO 60 I=1,NT
          X = X + (ARES(I)-ACLC(I))**2 + (BRES(I)-BCLC(I))**2
   60     CONTINUE
*                                            +++++ store r.m.s.e.
          ERR(I1) = SQRT(X)/ASQ
   50   CONTINUE
*                                            Proc.5.6. print errors
        WRITE(6,630)  PMW(NP2),(ERR(I1), I1=1,N1+1)
   40 CONTINUE

      RETURN

*****************  F O R M A T  ***********************************
  600 FORMAT(//1H0,'***** CALCULATION OF ROOT-MEAN-SQUARE-ERRORS *****')
  610 FORMAT( 1H0, 12X, 11E10.3 )
  620 FORMAT( 1H , 12X, 11A10 )
  630 FORMAT( 1H , E11.4, 11F10.5 )

      END
```

```
**************************************************************** ******
*                                                                    *
*      SUBROUTINE    RDVARI(NP1,N1,VS1,DP1, NP2,N2,VS2,DP2)          *
*                                                                    *
*      Proc.5.1.  set two variable parameters (data on horizontal & *
*                 vertical parameters are read from file '4' )       *
*                                                                    *
*      source       : SUB. RMSERR                                    *
*      routine used : none                                           *
*      data used    : none                                           *
*      returned data: NP1,N1,VS1,DP1,NP2,N2,VS2,DP2                  *
*          NP1,NP2 : integer;    parameter numbers                   *
*          N1, N2  : real;       number of divisions                 *
*          VS1,VS2 : real;       initial values                      *
*          VE1,VE2 : real;       final values                        *
*          (---1 : horizontal parameter, ---2 : vertical parameter)  *
*      format of input data from file '4' :                          *
*          NP1,N1,VS1,VE1: FORMAT( 2I5,F15.0,F10.0 )                 *
*          NP2,N2,VS2,VE2: FORMAT( 2I5,F15.0,F10.0 )                 *
*                                                                    *
**********************************************************************

      SUBROUTINE    RDVARI(NP1,N1,VS1,DP1, NP2,N2,VS2,DP2)
*                                             +++++ read two parametr
      READ(4,400)    NP1,N1,VS1,VE1
      WRITE(6,600)  NP1,N1,VS1,VE1
      READ(4,400)    NP2,N2,VS2,VE2
      WRITE(6,610)  NP2,N2,VS2,VE2
*                                             +++++ check parm values
       IF( NP1.LT.1 .OR. NP1.GT.20 )  GO TO 90
       IF( NP2.LT.1 .OR. NP2.GT.20 )  GO TO 90
       IF( N1.LT.0 )  THEN
          N1 = 0
          WRITE(6,620)  N1
        ELSE IF( N1.GT.10 )  THEN
          N1 = 10
          WRITE(6,620)  N1
       END IF
       IF( N2.LT.0 )  THEN
          N2 = 0
          WRITE(6,630)  N2
        ELSE IF( N2.GT.10 )  THEN
          N2 = 10
          WRITE(6,630)  N2
       END IF
*                                             +++++ calc increments
       IF( N1.EQ.0 )  THEN
          DP1 = 0
        ELSE
          DP1 = (VE1-VS1)/FLOAT(N1)
       END IF
       IF( N2.EQ.0 )  THEN
          DP2 = 0
        ELSE
          DP2 = (VE2-VS2)/FLOAT(N2)
       END IF
```

```
         RETURN
*                                                    print error message
     90  WRITE(6,640)
         STOP

********************  F O R M A T  ***************************************
    400 FORMAT( 2I5, F15.0, F10.0 )
    600 FORMAT( 1H0, 5X, 'HORIZONTAL PARAM. NO.', I2, 5X, 'NO OF DIVISION
       $=', I3, 3X, '(', E10.3, ' TO', E10.3, ')' )
    610 FORMAT( 1H0, 5X, 'VERTICAL   PARAM. NO.', I2, 5X, 'NO OF DIVISION
       $=', I3, 3X, '(', E10.3, ' TO', E10.3, ')' )
    620 FORMAT( 1H0,'NUMBER OF DIVISION (HORIZONTAL PARAMETER) IS REPLACED
       $ BY', I5 / )
    630 FORMAT( 1H0, 'NUMBER OF DIVISION (VERTICAL PARAMETER) IS REPLACED
       $BY', I5 / )
    640 FORMAT( 1H0, 'PARAMETER NUMBER IS NOT ADEQUATE. (SUB. RDVARI)' / )

         END
```

B.2   Data Input

The measured input and response signals are stored in file '1'; the half period and number of terms for Fourier expansion are in file '2'; the packed-bed parameters are in file '3'; control parameters for construction of two-dimensional error map are in file '4'. The following illustration is based on the given data of Run 3 in Example 1.2:

file '1': measured input and response signals

ADSORPTION GAS CHROMATOGRAPHY DATA   RUN 3   (RE = 0.30)

0.0	0.3125								
0.0	6.5	30.5	81.5	133.0	184.0	225.5	251.0	270.5	281.0
282.5	273.5	257.0	238.0	207.0	186.0	158.5	138.5	112.0	99.0
83.5	73.5	60.0	52.5	45.0	40.0	34.5	31.5	27.5	24.0
22.0	19.5	17.5	16.0	14.5	13.0	12.0	11.0	10.5	9.5
9.0	8.0	7.5	7.0	6.5	6.0	5.5	5.5	5.0	5.0
4.5	4.5	4.0	4.0	3.5	3.5	3.0	3.0	2.5	2.5
2.0	2.0	1.5	1.5	1.0	1.0	1.0	1.0	0.5	0.5
0.5	0.5	0.5	0.5	0.5	0.5	0.5	0.5	0.5	0.5
0.5	0.5	0.5	0.0	−1.0					
23.75	1.25								
0.0	0.6	0.8	1.1	1.3	1.7	2.2	3.2	4.7	8.3
12.3	17.3	23.3	30.9	40.4	50.4	61.0	72.5	83.0	93.0
100.1	107.1	111.1	114.1	115.2	114.2	111.2	106.7	101.3	94.3
87.3	80.4	72.4	64.4	56.9	50.5	43.0	37.5	32.0	27.1
22.6	19.1	16.2	13.7	11.2	8.7	6.8	5.3	4.3	3.3
2.4	1.4	0.9	0.4	0.0	−1.0				

file '2': parameters for Fourier expansion

70.0	50

file '3': packed-bed parameters

0.204	0.042	1860.	0.38	0.001	0.59	1.0	0.76E-4
0.48E-4	0.63E-6	2.6	0	5.29			

file '4': vertical and horizontal parameters for construction of error map

9	5	0.4E-4	0.5E-4
13	10	5.5	5.0

Appendix C. Derivations of Moment-equations in digital computer

C.1   Derivations of Eqs. (1.63) and (1.64)

Using the relations given by Eqs. (1.18) and (1.19), Eqs. (1.63) and (1.64) are derived from the transfer function of Eq. (1.61). A REDUCE 2† program for the derivations together with the obtained moment-equations are listed below.

```
COMMENT   ****************************************************************
          *                                                            *
          *     MOMENT EQUATIONS FOR INFINITE ADSORPTION BEDS          *
          *                                                            *
          *       DAX : axial dispersion coefficient                   *
          *       DE  : intraparticle effective diffusivity            *
          *       EB  : bed void fraction                              *
          *       EP  : intraparticle void fraction                    *
          *       KA  : adsorption equilibrium constant                *
          *       L   : distance between measuring points              *
          *       R   : particle radius                                *
          *       RHOP: particle density                               *
          *       S   : Laplace operator                               *
          *       SA  : surface area per unit volume of packed bed     *
          *       SKA : adsorption rate constant                       *
          *       SKF : particle-to-fluid transfer coefficient         *
          *       U   : interstitial flow rate                         *
          ****************************************************************;

COMMENT   ****************************************************************
          *     LAMB(SQ) : logarithm of transfer function              *
          *       SQ  = S+Q(P(S)),    Q(P) : eqn(1.61b)                *
          *       P(S)= (PHI-A)**2, PHI-A : eqn(1.61c)                 *
          ****************************************************************;

          PROCEDURE  LAMB(SQ);
           (L*U/(2*DAX))*(1-(1+4*DAX*SQ/U**2)**(1/2));

COMMENT   ****************************************************************
          *     EXP(X,N) : polynomial expression of E**X               *
          *       N : highest order                                    *
          ****************************************************************;

          PROCEDURE  EXP(X,N);
           1 + (FOR I:=1:N SUM X**I/(FOR J:=1:I PRODUCT J));

COMMENT   ****************************************************************
          *     Q(P,N) : polynomial expression of Q(P)                 *
          *       Q(P) : eqn(1.61b), P = (PHI-A)**2, N : highest order *
          ****************************************************************;
```

† A. C. Hearn, UCP-19, Univ. of Utah, March 1973.

346

```
PROCEDURE  Q(P,N);
  BEGIN  INTEGER M;
    M := 2*N$
    LET  X**2 = P;
    COTH := (EXP(X,M)+EXP(-X,M))/(EXP(X,M+1)-EXP(-X,M+1))$
    RETURN  (SA*DE/(R*EB))/(DE/(R*SKF)+1/(X*COTH-1));
  END;
```

```
COMMENT   ************************************************************
          *       P(S) : PHI-A squared                              *
          ************************************************************;
```

```
PROCEDURE  P(S);
  R**2*(S/DE)*(EP+RHOP*SKA*KA/(KA*S+SKA));
```

```
COMMENT   ************************************************************
          *       derivation of first moment                       *
          *          first moment : F1 = -L1*(1+R1), R1 = Q1*P1     *
          ************************************************************;
```

```
    L1 := SUB(SQ=0,DF(LAMB(SQ),SQ))$
    Q1 := SUB(Y=0,DF(Q(Y,1),Y))$
    P1 := SUB(S=0,DF(P(S),S))$
    R1 := Q1*P1$

OFF EXP;
    F1 := -L1*(1+R1)$
ON DIV;
    WRITE "FIRST MOMENT = ", F1;
```

```
COMMENT   ************************************************************
          *       derivation of second central moment              *
          *          second central moment : F2 = L2*(1+R1)**2      *
          *                                    + L1*(Q2*P1**2+Q1*P2) *
          ************************************************************;
```

```
ON EXP;
    L2 := SUB(SQ=0,DF(LAMB(SQ),SQ,2))$
    Q2 := SUB(Y=0,DF(Q(Y,2),Y,2))$
    P2 := SUB(S=0,DF(P(S),S,2))$

OFF EXP;
    F2 := L2*(1+R1)**2 + L1*(Q2*P1**2+Q1*P2)$
    WRITE "SECOND CENTRAL MOMENT = ", F2;

END;
```

Computer output:

$$\text{FIRST MOMENT} = 1/3 * EB^{(-1)} * L * U^{(-1)} * (3 * EB + EP * R * SA + KA * R * RHOP * SA)$$

$$\begin{aligned}
\text{SECOND CENTRAL MOMENT} = {}& 2/9 * DAX * EB^{(-2)} * L * U^{(-3)} * (3 * EB \\
& + EP * R * SA + KA * R * RHOP * SA)^2 \\
& - 1/45 * DE^{(-1)} * EB^{(-1)} * L * R^{(-2)} * SKA^{(-1)} \\
& * SKF^{(-1)} * U^{(-1)} * (-30 * DE * KA^2 * R^3 * RHOP \\
& * SA * SKF + SKA * (-10 * DE * SA - 2 * R * SA \\
& * SKF) * (EP * R^2 + KA * R^2 * RHOP)^2)
\end{aligned}$$

C.2   Derivation of First Moment of the System Discussed in Section 1.3

Laplace transformations of Eqs. (1.77) and (1.78) yield (also refer to Figure 1.24)

$$\bar{C}_i = A_i \exp\left[\frac{\lambda_D x}{L_D}\right] + B_i \exp\left[\frac{(\lambda_D + \sigma_D) x}{L_D}\right] \tag{C1}$$

where

$$0 < x < L_D \qquad\qquad \text{for Section } i = 1$$
$$L + L_D < x < L + 3L_D \qquad\qquad \text{for Section } i = 3$$

and

$$\bar{C}_i = A_i \exp\left[\frac{\lambda_B x}{L}\right] + B_i \exp\left[\frac{(\lambda_B + \sigma_B) x}{L}\right] \tag{C2}$$

where

$$L_D < x < L + L_D \qquad\qquad \text{for Section } i = 2$$
$$x > L + 3L_D, \text{ and } B_4 = 0 \qquad\qquad \text{for Section } i = 4$$

Consider the following two-component vectors:

$$\bar{C}_i = \begin{bmatrix} \bar{C}_i \\ D' \dfrac{d\bar{C}_i}{dx} \end{bmatrix} \qquad\qquad \text{for } i = 1 \text{ and } 3 \tag{C3}$$

and

$$\bar{C}_i = \begin{bmatrix} \bar{C}_i \\ \epsilon_b D_{ax} \dfrac{d\bar{C}_i}{dx} \end{bmatrix} \qquad\qquad \text{for } i = 2 \text{ and } 4 \tag{C4}$$

The boundary conditions are then expressed as

$$\bar{C}_i = \bar{C}_{i+1} \qquad \text{at } x = L_D, L + L_D \text{ and } L + 3L_D \tag{C5}$$

Also, from Eqs. (C1)–(C4) the following equations are derived:

$$(\bar{C}_1)_{x=0} = M\left(\lambda_D, \lambda_D + \sigma_D, \frac{D'}{L_D}\right) \exp\left[-\lambda_D\right] (\bar{C}_1)_{x=L_D} \tag{C6}$$

$$(\bar{C}_2)_{x=L_D} = M\left(\lambda_B, \lambda_B + \sigma_B, \frac{\epsilon_b D_{ax}}{L}\right) \exp\left[-\lambda_B\right] (\bar{C}_2)_{x=L+L_D} \tag{C7}$$

$$(\bar{C}_3)_{x=L+L_D} = M\left(2\lambda_D, 2\lambda_D + 2\sigma_D, \frac{D'}{2L_D}\right) \exp\left[-2\lambda_D\right] (\bar{C}_3)_{x=L+3L_D} \tag{C8}$$

$$(\bar{C}_3)_{x=L+2L_D} = M\left(\lambda_D, \lambda_D + \sigma_D, \frac{D'}{L_D}\right) \exp\left[-\lambda_D\right] (\bar{C}_3)_{x=L+3L_D} \tag{C9}$$

and

$$(\bar{C}_4)_{x=L+3L_D} = A_4 \exp\left[\lambda_B\left(1 + \frac{3L_D}{L}\right)\right]\begin{bmatrix} 1 \\ \dfrac{\epsilon_b D_{ax}\lambda_B}{L} \end{bmatrix} \tag{C10}$$

where

$$M(\alpha, \beta, \gamma) = \frac{1}{\beta-\alpha}\begin{bmatrix} \beta-\alpha\exp[\alpha-\beta] & -\dfrac{1}{\gamma}(1-\exp[\alpha-\beta]) \\ \alpha\beta\gamma(1-\exp[\alpha-\beta]) & -(\alpha-\beta\exp[\alpha-\beta]) \end{bmatrix}$$

Transfer function of the system between $x=0$ and $x=L+2L_D$ is then

$$F(s) = \frac{(\bar{C}_3)_{x=L+2L_D}}{(\bar{C}_1)_{x=0}}$$

$$= \frac{[1\ \ 0]M\left(\lambda_D, \lambda_D+\sigma_D, \dfrac{D'}{L_D}\right)\begin{bmatrix} 1 \\ \dfrac{\epsilon_b D_{ax}\lambda_B}{L} \end{bmatrix}\exp[\lambda_B+2\lambda_D]}{[1\ \ 0]M\left(\lambda_D, \lambda_D+\sigma_D, \dfrac{D'}{L_D}\right)M\left(\lambda_B, \lambda_B+\sigma_B, \dfrac{\epsilon_b D_{ax}}{L}\right)}$$

$$\times M\left(2\lambda_D, 2\lambda_D+2\sigma_D, \dfrac{D'}{2L_D}\right)\begin{bmatrix} 1 \\ \dfrac{\epsilon_b D_{ax}\lambda_B}{L} \end{bmatrix} \tag{C11}$$

A REDECE 2 program for the derivation of first moment according to Eq. (1.18) is listed together with the obtained moment-equation below. (Note that in the program the matrix elements are expressed as linear function of s. This is good enough for the derivation of first moment.)

```
COMMENT   *************************************************************
          *                                                           *
          *      MOMENT EQUATION FOR ADSORPTION/DEAD VOLUME SYSTEM     *
          *                                                           *
          *************************************************************;
COMMENT   *************************************************************
          *    M(I,J,AL,BE,GAM)  : matrix's (I,J)-component expressed  *
          *                        as linear function of S            *
          *************************************************************;
          PROCEDURE  M(I,J,AL,BE,GAM);
            BEGIN;
              IF I=1 AND J=1 THEN  MX := (BE-AL*E**(AL-BE))/(BE-AL)$
              IF I=1 AND J=2 THEN  MX :=-(1/GAM)*(1-E**(AL-BE))/(BE-AL)$
              IF I=2 AND J=1 THEN  MX := AL*BE*GAM*
                                         (1-E**(AL-BE))/(BE-AL)$
              IF I=2 AND J=2 THEN  MX :=-(AL-BE*E**(AL-BE))/(BE-AL)$
              RETURN  SUB(S=0,MX)+SUB(S=0,DF(MX,S))*S;
            END;
```

```
COMMENT  ***********************************************************
         *       derivation of first moment                        *
         *       DAX : axial dispersion coefficient                 *
         *       DD  : dispersion coefficient in dead volume section *
         *       EB  : bed void fraction                            *
         *       L   : packed bed length                            *
         *       LD  : half length of dead volume section           *
         *       S   : Laplace operator                             *
         *       U   : interstitial fluid velocity                  *
         *       UD  : fluid velocity in dead volume section        *
         ***********************************************************;

COMMENT  ***********************************************************
         *       parameters as linear function of S                 *
         *       LAMB : eqn(1.80d), LAMD : eqn(1.80e)               *
         *       SIGB : eqn(1.80f), SIGD : eqn(1.80g)               *
         *       AL = EB*DAX/DD, BE = EB*U*LD/DD                     *
         *       Q1 = 1+DEL0, DEL0 : eqn(1.63a)                      *
         ***********************************************************;

         LET  UD = EB*U,
              EB*DAX = AL*DD,
              EB*U*LD = BE*DD;
         LAMB :=-(L/U)*Q1*S$
         LAMD :=-(LD/UD)*S$
         SIGB := U*L/DAX + 2*(L/U)*Q1*S$
         SIGD := UD*LD/DD + 2*(LD/UD)*S$

COMMENT  ***********************************************************
         *       matrices and vectors in eqn(C11)                   *
         ***********************************************************;

         MATRIX  M1(2,2),M2(2,2),M3(2,2),V1(1,2),V2(2,1);
         V1 := MAT((1,0))$
         V2 := MAT((1),(EB*DAX*LAMB/L))$
         FOR I:=1:2  DO BEGIN;
           FOR J:=1:2  DO BEGIN;
             M1(I,J) := M(I,J,LAMD,LAMD+SIGD,DD/LD)$
             M2(I,J) := M(I,J,LAMB,LAMB+SIGB,EB*DAX/L)$
             M3(I,J) := M(I,J,2*LAMD,2*(LAMD+SIGD),DD/(2*LD))$
           END;
         END;

COMMENT  ***********************************************************
         *       transfer function : (FNUM/FDEN)*E**FEXP            *
         ***********************************************************;

         FNUM := V1*M1*V2$
         FDEN := V1*M1*M2*M3*V2$
         FEXP := LAMB+2*LAMD$

COMMENT  ***********************************************************
         *       derivation of first moment, F1 : first moment      *
         ***********************************************************;

         F1 :=  -SUB(S=0,DF(FNUM,S))/SUB(S=0,FNUM)
                + SUB(S=0,DF(FDEN,S))/SUB(S=0,FDEN)
                - SUB(S=0,DF(FEXP,S))$

         ON DIV;
           WRITE "FIRST MOMENT = ", F1;

         END;
```

Computer output:

$$\begin{aligned}
\text{FIRST MOMENT} = {}& E^{((-BE*DAX-L*U)/DAX)}*AL*DD*EB^{(-1)}*Q1*U^{(-2)} \\
& - E^{((-3*BE*DAX-L*U)/DAX)}*AL*DD*EB^{(-1)}*Q1*U^{(-2)} \\
& + 2*BE*DD*EB^{(-2)}*U^{(-2)} - E^{((-BE*DAX-L*U)/DAX)} \\
& *DD*EB^{(-2)}*U^{(-2)} + E^{((-3*BE*DAX-L*U)/DAX)} \\
& *DD*EB^{(-2)}*U^{(-2)} + L*Q1*U^{(-1)}
\end{aligned}$$

Author Index

Subject Index